Discourse on Method *and* Meditations on First Philosophy

RENÉ DESCARTES

Edited and Introduced by

ANDREW BAILEY

Translated by

IAN JOHNSTON

broadview press

BROADVIEW PRESS – www.broadviewpress.com
Peterborough, Ontario, Canada

Founded in 1985, Broadview Press remains a wholly independent publishing house. Broadview's focus is on academic publishing; our titles are accessible to university and college students as well as scholars and general readers. With 800 titles in print, Broadview has become a leading international publisher in the humanities, with world-wide distribution. Broadview is committed to environmentally responsible publishing and fair business practices.

Library and Archives Canada Cataloguing in Publication

Title: Discourse on method and Meditations on first philosophy / René Descartes ; edited and introduced by Andrew Bailey ; translated by Ian Johnston.
Names: Descartes, René, 1596-1650, author. | Bailey, Andrew, 1969- editor. | Johnston, Ian (Ian Courtenay), 1938- translator. | Container of (expression): Descartes, René, 1596-1650. Discours de la méthode. English. | Container of (expression): Descartes, René, 1596-1650. Meditationes de prima philosophia. English.
Description: Translations of Discours de la méthode and Meditationes de prima philosophia. | Includes bibliographical references and index.
Identifiers: Canadiana (print) 20200285815 | Canadiana (ebook) 20200285955 | ISBN 9781554815548 (softcover) | ISBN 9781770487840 (PDF) | ISBN 9781460407356 (EPUB)
Subjects: LCSH: Methodology. | LCSH: Knowledge, Theory of. | LCSH: Science—Methodology. | LCSH: First philosophy.
Classification: LCC B1848.E5 J64 2020 | DDC 194—dc23

Broadview Press handles its own distribution in North America:
PO Box 1243, Peterborough, Ontario K9J 7H5, Canada
555 Riverwalk Parkway, Tonawanda, NY 14150, USA
Tel: (705) 743-8990; Fax: (705) 743-8353
email: customerservice@broadviewpress.com

For all territories outside of North America, distribution is handled by Eurospan Group.

Broadview Press acknowledges the financial support of the Government of Canada for our publishing activities.

Canada

Copy-edited by Robert M. Martin and Michel Pharand
Typeset by Eileen Eckert
Cover design by Lisa Brawn

PRINTED IN CANADA

Discourse on Method
and
Meditations on First Philosophy

CONTENTS

INTRODUCTION TO DESCARTES

Who Was René Descartes?

René Descartes was born in 1596 in a small town nestled below the vineyards of the Loire in western France; at that time the town was called La Haye, but it was later renamed Descartes in his honor. His early life was probably unhappy: he suffered from ill health, his mother died a year after he was born, and he didn't get on well with his father. (When René sent his father a copy of his first published book, his father's only reported reaction was that he was displeased to have a son "idiotic enough to have himself bound in vellum."[1]) At the age of about 10 he went to the newly founded college of La Flèche to be educated by the Jesuits. Descartes later called this college "one of the best schools in Europe," and it was there that he learned the medieval "scholastic" science and philosophy that he was later decisively to reject. Descartes took a law degree at the University

1 Vellum is the parchment made from animal skin that was used to make books.

of Poitiers and studied mathematics and mechanics; then, at 21, he joined first the Dutch army of Prince Maurice of Nassau and then the forces of Maximilian of Bavaria. As a soldier he saw little action, traveling around Europe supported by his family's wealth.

During this period, he resolved "to stop seeking any other science except one which could be found inside myself or in the great book of the world," developing an intense interest in mathematics, which stayed with him for the rest of his life. In fact, Descartes was one of the most important figures in the development of algebra, which is the branch of mathematics that allows abstract relations to be described without using specific numbers, and which is therefore capable of unifying arithmetic and geometry:[2]

> I came to see that the exclusive concern of mathematics is with questions of order or method, and that it is irrelevant whether the measure in question involves numbers, shapes, stars, sounds, or any other object whatsoever. This made me realize that there must be a general science which explains all the points that can be raised concerning order and measure irrespective of subject matter. (from *Rules for the Direction of Our Native Intelligence* [1628])

This insight led Descartes directly to one of the most significant intellectual innovations of the modern age: the conception of science as the exploration of abstract mathematical descriptions of the world.

It was also during this time—in 1619—that Descartes had the experience said to have inspired him to take up the life of a philosopher, and which, perhaps, eventually resulted in the form of the *Meditations*. Stranded by bad weather near Ulm on the river Danube, Descartes spent the day in a *poêle* (a stove-heated room[3]) engaged in intense philosophical speculations. That night he had three vivid dreams which he later described as giving him his mission in life. In the first dream Descartes felt himself attacked by phantoms and then a great wind; he was then greeted by a friend who gave him a message about a gift. On awaking after this first dream, Descartes felt a sharp pain which made him fear that the dream was the work of some deceitful evil demon. Descartes eventually fell back

---

2 He invented the method still used to quantify locations on a graph: Cartesian coordinates (that adjective is derived from the Latin version of his name, 'Cartesius').

3 Sometimes Descartes's words are taken in what might be their literal meaning, that he spent time in a stove. Although there is other evidence of his eccentricity, this seems an uncharitable translation. Often his phrase is taken to mean a stove-heated room; sometimes a small compartment attached to a masonry stove, with bedding, used for sleeping in cold weather.

asleep and immediately had the second dream: a loud thunderclap, which woke him in terror believing that the room was filled with fiery sparks. The third and last dream was a pleasant one, in which he found an encyclopedia on a table next to a poetry anthology, open to a poem which begins with the line "Which road in life shall I follow?" A man then appeared and said "*Est et non*"—"it is and is not." While still asleep, Descartes apparently began to speculate about the meaning of his dreams and decided, among other things, that the gift of which his friend spoke in the first dream was the gift of solitude, the dictionary represented systematic knowledge, and "*Est et non*" spoke of the distinction between truth and falsity as revealed by the correct scientific method. Descartes concluded that he had a divine mission to found a new philosophical system to underpin all human knowledge.

In 1628, at the age of 32, Descartes settled in the Netherlands (at the time the most intellectually vibrant region of Europe), where he lived for most of his remaining life. It was only then that he began sustained work in metaphysics and mathematical physics. His family was wealthy enough that Descartes, who cultivated very modest tastes, was free of the necessity to earn a living and could devote his time to scientific experimentation and writing. By 1633 he had prepared a book on cosmology and physics, called *Le Monde* (*The World*), in which he accepted Galileo's revolutionary claim that the Earth orbits the sun (rather than the other way around), but when he heard that Galileo had been condemned by the Inquisition of the Catholic Church, Descartes withdrew the work from publication.[4] In 1637 he published (in French) a sample of his scientific work, *Optics*, *Meteorology*, and *Geometry*, together with the *Discourse on the Method for Reasoning Well and for Seeking Truth in the Sciences*. Criticisms of this methodology led Descartes to write *Meditations on First Philosophy* in 1641. In 1644 he published a summary of his scientific and philosophical views, the *Principles of Philosophy*, which he hoped would become a standard university textbook, replacing the medieval texts used at the time. His last work, published in 1649, was *The Passions of the Soul*, which attempted to extend his scientific methodology to ethics and psychology.

Descartes never married, but in 1635 he had a daughter, Francine, with a serving woman called Hélène Jans. He made arrangements to care for and educate the girl but she died of scarlet fever at the age of five, a devastating shock for Descartes.

---

4 Descartes was very aware of the Catholic authorities' opposition to his ideas, and afraid of it. After his death, the Church placed all his works on the Index of Prohibited Works, with the note that they would remain there "until corrected." (The Church announced in 1966 that the prohibition of items on this list was no longer to be considered law, but the Index was retained as a moral guide, with Descartes and many other famous philosophers still on it.)

In 1649 Descartes accepted an invitation to visit Stockholm and give philosophical instruction to Queen Christina of Sweden. He was required to give tutorials at the royal palace at five o'clock in the morning. Ever since he was a sickly schoolboy, he stayed in bed until 11 a.m., and it is said that the strain of this sudden break in his habits and the harsh Swedish winter caused him to catch pneumonia; he died in February 1650. His dying words are said to have been, "*mon âme, il faut partir*"—"my soul, it's time we must leave." His body was returned to France but, apparently, his head was secretly kept in Sweden; in the 1820s a skull bearing the faded inscription "René Descartes" was discovered in Stockholm and is now on display in the Museum of Natural History in Paris.

What Was Descartes's Overall Philosophical Project?

Descartes lived at a time when the accumulated beliefs of centuries—assumptions based on religious doctrine, straightforward observation, and common sense—were being gradually but remorselessly stripped away by exciting new discoveries. (The most striking example of this was the evidence mounting against the centuries-old belief that an unmoving Earth is the center of the universe, orbited by the moon, sun, stars, and all the other planets.) In this intellectual climate, Descartes became obsessed by the thought that no lasting scientific progress was possible without a systematic method for sifting through our preconceived assumptions and distinguishing between those that are reliable and those that are false. Descartes's central intellectual goal was to develop just such a reliable scientific method, and then to construct a coherent and unified theory of the world and of humankind's place within it. This theory, he hoped, would replace scholasticism, the deeply flawed medieval system of thought based on the science of Aristotle and Christian theology.

A key feature of Descartes's system is that all knowledge should be based on utterly reliable foundations, discovered through the systematic rejection of any assumptions that can possibly be called into doubt. Then, as in mathematics, complex conclusions could be reliably derived from these foundations by chains of valid reasoning—of simple and certain inferences. The human faculty of *reason* was therefore of the greatest importance. Furthermore, Descartes urged that scientific knowledge of the external world should be rooted, not in the deceptive and variable testimony of the senses, but in the concepts of pure mathematics. That is, Cartesian science ("Cartesian" being the adjective derived from Descartes's name) tries to reduce all physics to "what the geometers call *quantity*, and take as the object of their demonstrations, i.e., that to which every kind of division, shape, and motion is applicable" (*Principles of Philosophy* 1644). There is, however, for Descartes, a place for empirical investigation in science—not as a tool for producing general understanding, but rather to determine the real external existences of particular things.

These ideas (though they have never been uncritically and uniformly accepted) have come to permeate the modern conception of science, including Descartes's influential metaphor of a unified "tree of knowledge," with metaphysics as the roots, physics as the trunk, and the special sciences (such as biology, anthropology, or ethics) as the branches. His most important and lasting influence on scientific thought is his idea that the physical world is a unified whole, governed by very basic universal mathematical and physical laws, and that finding these is the most fundamental job of science.

One much less familiar, and less enduring, aspect of Descartes's method for the production of knowledge is the central role played by God in his system. For Descartes, all human knowledge of the world around us essentially relies upon our prior knowledge that a non-deceiving God exists. Science, properly understood, not only does not conflict with religion but actually *depends* on religion, he believed.

Lastly, one of the best-known results of Descartes's metaphysical reflections is "Cartesian dualism." This doctrine holds that mind and body are two completely different substances—that the mind is a nonphysical self in which all the operations of thought take place, and which can be wholly separated from the body after death. Like much of Descartes's work, this theory achieved the status of a more or less standard view for some 300 years after his death, but at the time it was a radical philosophical innovation, breaking with the traditional Aristotelian conception of mental activity as a kind of *attribute* of the physical body (rather than as something entirely separable from the body).

How Important and Influential Are These Texts?

Descartes is one of the most widely studied of any of the Western philosophers. The *Discourse on Method* was his first major philosophical publication. It offers an outline of the intellectual and scientific method that he advocated and employed, as well as an early presentation of several of his key philosophical arguments.

The *Meditations on First Philosophy* is Descartes's philosophical masterpiece and arguably his most important work. John Cottingham, an expert on Descartes, has written of the *Meditations* that

> The radical critique of preconceived opinions or prejudices which begins that work seems to symbolize the very essence of philosophical inquiry. And the task of finding secure foundations for human

knowledge, a reliable basis for science and ethics, encapsulates, for many, what makes philosophy worth doing.[5]

"I think; therefore, I am" is the most famous dictum in the history of philosophy. Note that this is not what it is often popularly taken to be: praise of the intellectual life as the real source of human identity. It's rather Descartes's foundational claim beginning his reconstruction of indubitable truth. Descartes's proofs for the existence of God, and his use of those proofs to establish a footing for other beliefs, are also hugely influential, even if the majority of contemporary philosophers find fault in those arguments.

The importance of Descartes's work to the history of thought is profound. He is commonly considered the first great philosopher of the modern era, since his work was central in sweeping away medieval scholasticism based on Aristotelian science and Christian theology and replacing it with the methods and questions that dominated philosophy until the twentieth century.[6] This change from scholastic to modern modes of thought was also crucial to the phenomenal growth of natural science and mathematics beginning in the seventeenth century. In recent years, however, it has been fashionable to blame Descartes for what have been seen as philosophical dead ends, and many of the assumptions which he built into philosophy have been questioned (this is one of the reasons why the philosophy of the second half of the twentieth century and beyond has been so exciting).

Timeline

1596	Born in La Haye, France.
1607–14	Studies at the Collège Royal Henry-Le-Grand.
1615–16	Studies law at the University of Poitiers.
1618	Joins the Dutch army of Prince Maurice of Nassau.

---

5 This appears in his Introduction to the *Cambridge Companion to Descartes* (1992).

6 Alan Gewirth went so far as to write, in 1970, "the history of twentieth-century philosophy ... consists in a series of reactions to Descartes's metaphysics. Examples of these reactions are Ryle's castigations of the Cartesian mind-body dualism, Sartre's and Hare's attacks on Cartesian intellectualism and intuitionism, Chomsky's support of Cartesian innatism, and the opposed views taken on Cartesian doubt by Russell and Husserl on the one hand and by Moore, Dewey, Austin, and the later Wittgenstein on the other" ("The Cartesian Circle Reconsidered," *The Journal of Philosophy* 67, 668–85).

1619	A series of dreams leads Descartes to conclude that he must found a new philosophical system grounded in precise methods resembling those of mathematics.
1620	Leaves the army.
1627–28	The Siege of La Rochelle, in which the royalist forces of Louis XIII defeat the Protestant Huguenot forces. Descartes visits the site of the siege in 1627.
1628	Descartes moves to Holland, where he begins sustained and productive work in metaphysics and mathematical physics.
1633	Galileo's *Dialogue Concerning the Two Chief World Systems* is banned by the Roman Inquisition and Galileo is placed under house arrest.
	Descartes completes *The World*, a work which applies the heliocentric worldview, but delays publication indefinitely upon hearing of Galileo's conviction.
1635	Descartes's daughter, Francine, is born.
1637	Publishes the *Discourse on Method* (full title: *Discourse on the Method for Reasoning Well and for Seeking Truth in the Sciences*), along with the *Optics*, *Meteorology*, and *Geometry*, in which his methods are applied. The *Geometry* included one of the first presentations of what came to be known as the "Cartesian coordinate system."
1640	Francine dies of scarlet fever.
1641	Publishes the *Meditations* (full title: *Meditations on First Philosophy, in which the existence of God and the difference between the human soul and body are demonstrated*).
1643	Begins corresponding with Princess Elisabeth of Bohemia, who among other things challenges Descartes's views on dualism.
1644	Publishes *Principles of Philosophy*, a summary of his scientific and philosophical views.
1649	Publishes *The Passions of the Soul*, a work on psychology and the emotions.
	Moves to Stockholm to tutor Queen Christina of Sweden.
1650	Dies of pneumonia, in Stockholm.
1662	*Treatise on Man* is published.
1664	*The World* is published.
1701	*Rules for the Direction of the Mind*, an incomplete treatise on method which Descartes abandoned in 1628, is published.

Suggestions for Further Reading

Discourse on Method was originally published alongside the *Optics*, *Meteorology*, and *Geometry*. Those works are less philosophical than the *Discourse*, but they were influential in other domains (especially mathematics) and illustrate the application of Descartes's methods to scientific inquiry. Paul J. Olscamp has published a complete translation of all four texts (Hackett, 2001).

Meditations on First Philosophy was originally published with an extensive set of objections from contemporary thinkers and replies by Descartes. Cambridge University Press has published Descartes's collected philosophical writings and letters in three volumes, translated by John Cottingham, Robert Stoothoff, Dugald Murdoch, and Anthony Kenny. (Volume II, which includes the *Meditations* and the complete *Objections and Replies*, was published in 1984; Volume I, which includes the *Discourse on Method*, was published in 1985, and the letters in 1991.)

The secondary literature on Descartes is vast, but here are a few starting points. An entertaining biography of Descartes is from A.C. Grayling, *Descartes: The Life and Times of a Genius* (Walker & Company, 2006); an even more thorough biography is *Descartes: An Intellectual Biography* by Stephen Gaukroger (Oxford University Press, 1995). Useful general introductions to Descartes's thought are John Cottingham's *Descartes* (Blackwell, 1991), Anthony Kenny's *Descartes: A Study of His Philosophy* (Random House, 1968), Bernard Williams' *Descartes: The Project of Pure Enquiry* (Penguin, 1978, reissued by Taylor & Francis in 2007), and Margaret Dauler Wilson's *Descartes* (Routledge, 1978); a second edition of Georges Dicker's *Descartes: An Analytical and Historical Introduction* (Oxford University Press) was released in 2013. There is a *Blackwell Guide to Descartes' Meditations* (Wiley-Blackwell, 2006), edited by Stephen Gaukroger, and Gary Hatfield has written a Routledge GuideBook to *Descartes and the Meditations* (Taylor and Francis, 2007). *The Cambridge Companion to Descartes* edited by John Cottingham (Cambridge University Press, 1992) is very helpful, and there is also a Blackwell *Companion to Descartes* (2010) edited by Janet Broughton and John Carriero, as well as *The Oxford Handbook of Descartes and Cartesianism* (Oxford University Press, 2019) edited by Steven Nadler, Tad M. Schmaltz, and Delphine Antoine-Mahut. Two good collections of articles on the *Meditations* are A.O. Rorty's *Essays on Descartes's Meditations* (University of California Press, 1986) and Vere Chappell's *Descartes's* Meditations*: Critical Essays* (Rowman and Littlefield, 1997). Several important essays are also contained in Willis Doney (ed.), *Descartes: A Collection of Critical Essays* (University of Notre Dame Press, 1967).

A few of the more influential (and sometimes controversial) books on Descartes are Harry Frankfurt's *Demons, Dreamers, and Madmen* (Bobbs-Merrill, 1970), E.M. Curley's *Descartes against the Sceptics* (Harvard University Press, 1978), Peter J. Markie's *Descartes's Gambit* (Cornell University Press, 1986), Richard Watson's *The Breakdown of Cartesian Metaphysics* (Humanities Press, 1987), Daniel Garber's *Descartes's Metaphysical Physics* (University of Chicago Press, 1992), Janet Broughton, *Descartes's Method of Doubt* (Princeton University Press, 2002), Marleen Rozemond's *Descartes's Dualism* (Harvard University Press, 2002), Desmond Clarke's *Descartes's Theory of Mind* (Oxford University Press, 2005), Tom Sorrell's *Descartes Reinvented* (Cambridge University Press, 2005), and Deborah Brown's *Descartes and the Passionate Mind* (Cambridge University Press, 2008). A well-known book that is tangentially about Descartes is *Descartes' Error: Emotion, Reason, and the Human Brain*, by Antonio Damasio (Penguin, 2005), while Richard Rorty's *Philosophy and the Mirror of Nature* (Princeton University Press, 1979) attributes a wide swath of modern intellectual failings to the pernicious influence of Descartes.

A representative handful of the papers published on Descartes (excluding those in the collections already mentioned) are: Elizabeth Anscombe, "The First Person," in Anscombe, *Metaphysics and the Philosophy of Mind* (Blackwell, 1981), 21–36; O.K. Bouwsma, "Descartes' Evil Genius," *Philosophical Review*, 58 (1949), 141–51; John Cottingham, "Cartesian Trialism," *Mind* 94 (1985), 218–30; Alan Gewirth, "The Cartesian Circle," *Philosophical Review* 50 (1941), 368–95; Jaako Hintikka, "Cogito ergo sum: Inference or Performative?" *Philosophical Review* 71 (1962), 3–32; Anthony Kenny, "Descartes the Dualist," *Ratio* 12 (1999), 114–27; Louis Loeb, "The Priority of Reason in Descartes," *Philosophical Review* 99 (1990), 3–43; Norman Malcolm, "Descartes' Proof that He Is Essentially a Non-Material Thing," in Malcolm, *Thought and Knowledge* (Cornell University Press, 1977), 58–84; Dugald Murdoch, "The Cartesian Circle," *Philosophical Review* 108 (1999), 221–44; Stephen Schiffer, "Descartes on His Essence," *Philosophical Review* 85 (1976), 21–43; and James Van Cleve, "Foundationalism, Epistemic Principles, and the Cartesian Circle," *Philosophical Review* 88 (1979), 55–91.

TRANSLATOR'S NOTE

René Descartes (1596–1650) published *Discourse on Method* in 1637 as part of a work containing sections on optics, geometry, and meteorology. The fourth section, the *Discourse*, outlined the basis for a new method of investigating knowledge.

The translation of *Meditations on First Philosophy* is based upon the first Latin edition of Descartes's *Meditations* (1641). I have incorporated most of the relatively few corrections made to that text in the second Latin edition (1642), none of which are particularly important. I have also inserted a number of additions made to the Latin text in the French edition (1647), which was supervised by Descartes, who approved of the result. These additions from the French edition are inserted here only where they help to clarify the meaning of the original Latin (for example, by clarifying Descartes's Latin pronouns). Other changes in the French text I have ignored. Words in square brackets are insertions and additions from the first French edition.

*These translation have been revised and edited by the publisher.

**Numbers in the margins refer to the page numbers of the text in *Oeuvres de Descartes*, edited by Charles Adam and Paul Tannery.

DISCOURSE ON METHOD

What Is the Structure of the *Discourse*?

The *Discourse on Method* was originally published alongside three other works: *Optics*, *Meteorology*, and *Geometry*. It functioned as a kind of preface or methodological introduction to those scientific treatises, though it can be profitably read on its own. The *Discourse* establishes both the method which Descartes applied in those works and the premises upon which they depend for their foundations. For Descartes, one cannot be confident in one's beliefs about a scientific subject without first establishing the more basic beliefs upon which knowledge of those subjects can be built.

The text is in part autobiographical. While traveling and observing the practices and beliefs of other cultures, Descartes came to realize that many of his own beliefs were in fact false, and that others were seemingly arbitrary products of his upbringing. In order to sort between true and false beliefs, Descartes found it necessary to adopt a high standard for knowledge—to "not accept anything as true which I did not clearly know to be true." By adopting this principle and other general rules to guide his thought, Descartes believed he could establish a firm footing that would guide him to truth and serve to clear out the arbitrary beliefs that had previously clouded his judgment. The first three parts of the *Discourse* lay out the foundations of this project, establishing the rules that will guide his inquiry and the practical maxims that he will adopt in the meantime.

In the fourth part, Descartes begins to establish some of the foundational beliefs that meet his high standards. The most famous of these consists in a proof of his own existence: "I think; therefore, I am." From there, Descartes offers arguments for the existence of God and the human soul, and for the reliability of reason when it is carefully applied. Many of these arguments move quickly, and some of them are articulated in greater detail or in more nuanced forms in the *Meditations on First Philosophy*.

One can read the *Discourse* as Descartes's account of his own thought processes as he came to doubt and rebuild his beliefs; indeed, the past tense adopted throughout the text invites this interpretation. However, the *Discourse* is also meant as an invitation to the reader, showing how any person can undertake the project of reforming their beliefs by adopting strict standards for knowledge and by grounding scientific beliefs in careful reasoning from premises that have been demonstrated with certainty.

Some Useful Background Information

1. Descartes was an important figure in the scientific revolution of the sixteenth and seventeenth centuries, along with Francis Bacon, Galileo Galilei, Robert Boyle, Isaac Newton, and others. He produced innovative works in a variety of scientific fields, perhaps most notably mathematics. But it is his methodology, especially that which is presented in the *Discourse*, which arguably constitutes his most important contribution to modern science.

   At the time of this work's publication, very little had been written to defend and describe what we might now recognize as the scientific method. Descartes's acknowledgment and advocacy of careful and methodical inquiry, and his insistence on not adopting beliefs until they had been firmly established, set standards for inquiry that were not common among proto-scientific thinkers of the early seventeenth century.

2. Descartes's metaphysics was a radical departure from the then-prevailing Aristotelian view of nature. In very brief summary, Aristotelians saw natural bodies as being composed of both form and matter. Matter cannot exist without form, and a thing's form determines its nature: that is, a thing's form makes it what it is—a horse, a tree, a cloud—by determining its characteristic development and behavior. There are four basic substances—earth, air, fire, and water—and four basic qualities—hot, cold, wet, and dry. Most natural bodies are made up of mixtures of elements, and they belong to different kinds (such as species or types of minerals) defined by their forms. In this way, Aristotelian science made no distinction between biological processes, such as growth or nutrition, and other natural phenomena such as burning or gravity: they are all the playing out of essential forms, principles of growth and change, that are 'built in' to the entities involved.

   For the Aristotelian, then, scientific explanation will be a matter of identifying the multifarious forms—the essential natures—of all the different items in the natural world. Furthermore, the objects that we encounter in nature really have the properties they appear to have to our senses—color, texture, taste, odor, and so on—and what happens when we perceive the world is that these qualities are transferred from external objects to our sense organs (where they are received in our sensory soul as a "form without matter"). One final important aspect of Aristotelian science: the changing, natural realm in which we live is located at the center of the universe (due to the tendency of earth and water, because of their natures, to seek the center and thus collect there), but there is a radical discontinuity between this world and the heavenly spheres—literally, crystalline spheres in which the moon, sun, planets, and stars were thought to be embedded—which are unchanging, and not even made of the familiar four elements but of a completely different fifth element (called quintessence).

One of the interpretive tasks in reading Descartes is to discover the differences between his approach and that of the Aristotelian tradition. Some of the Cartesian departures to look for are: that different natural kinds differ only in the sizes, shapes, and motions of the particles that make them up; that matter does not contain within it its own principle of motion and change but is 'passive' and subject to external forces; that the qualities we encounter in sense experience do not resemble the causes of those experiences; and that the whole material world, including the heliocentric solar system, is governed by the same small set of laws of motion.

3. One of the distinctive aspects of Descartes's method, and one of the ways in which his method departs from scientists and philosophers who preceded him, is its mechanistic approach to the physical world. In Part 5 of the *Discourse*, Descartes discusses the heart and the circulatory system to a level of detail that may be surprising in the context of a largely philosophical text.[1] This discussion serves to demonstrate a method of reasoning in which rigorous empirical observations[2] are used in conjunction with logical inferences to develop a clearer picture of the world. But it also presents the body as a kind of machine, far more complex than the "moving machines" that people had created even in Descartes's day, but not different in kind. To interpret natural entities this way—as a collection of causal interactions that are governed by stable laws—is to believe that we can develop a purely mechanical understanding of the world. This idea is in contrast to earlier Aristotelian understandings of science, according to which natural entities strive toward a certain *telos* or purpose. In a classic example from Aristotle, it is said that an acorn's potential is to eventually become an oak tree; to know this potential

1 In addition to providing an example for what we consider Descartes's more "philosophical" ideas, this sort of thing, which we would today distinguish as science, was until much later considered part of philosophy—"*natural* philosophy." Newton's major work (1687) was titled *Philosophiæ Naturalis Principia Mathematica* (*Mathematical Principles of Natural Philosophy*).

2 An empirical observation is one that is based in sensory experience (as opposed to rational reflection).

is to understand the acorn. Descartes's mechanistic approach, on the other hand, would require that one investigate the causal processes at work: why the acorn sprouts, what causes it to grow, and what functions its various parts serve in those processes. Modern science is grounded in the mechanistic approach to science defended by Descartes and other early advocates such as Thomas Hobbes.

Some Common Misconceptions

1. The "method of doubt" is not an everyday method—it is not supposed to be an appropriate technique for making day-to-day decisions, or even for doing science or mathematics. Most of the time it would be hugely impractical for us to call into question everything that we might possibly doubt, to question all our presuppositions, before we make a judgment. Instead, the method of doubt is supposed to be a once-in-a-lifetime exercise, by which we discover and justify the basic "first principles" that we rely on in everyday knowledge. In short, we must always rely on certain assumptions when we make decisions or do science, and this is unavoidable but dangerous. Even when engaging in the method of doubt and reconstructing his belief system, Descartes adopts a set of principles comprising a "provisional morality," so as to have a practical (if uncertain) basis for guiding his actions.

2. It is sometimes supposed that Descartes thought that all knowledge could be mathematically deduced from the foundational beliefs that remain after he has applied his method of doubt. But this is not quite right. He thought that the proper concepts and terms, which science must use to describe the world, were purely mathematical and were discoverable through rational reflection alone. But he also recognized that only through empirical investigation can we discover which scientific descriptions, expressed in the proper mathematical terms, are actually true of the world. For example, reason tells us (according to Descartes—and this was a radically new idea at the time) that matter can be defined simply in terms of extension in three dimensions, and that the laws which guide the movements of particles can be understood mathematically. However, only experience can tell us how the bits of matter in, for example, the human body are actually arranged.

3. Descartes does not conclude that error is impossible, even for those who adopt the proper intellectual methods of science. He argues only that *radical and systematic* error is impossible for the conscien-

tious thinker. For example, even after working through Descartes's reasoning we might still occasionally be tricked by perceptual illusions, or think we are awake when in fact we are dreaming; what Descartes thinks he has blocked, however, is the possibility that such errors show that our entire picture of reality might be wrong.

Suggestions for Critical Reflection

1. In Part 2 of the *Discourse*, Descartes lays out four rules that apparently summarize his method for the successful pursuit of knowledge. To what extent do these rules provide us with a complete formula or recipe for doing science (or acquiring reliable knowledge more generally)? To what extent do they correspond to your understanding of the modern 'scientific method'? Does Descartes actually intend these four rules to be a complete encapsulation of his method, or are they limited to establishing the foundations that a method will need to build upon?

2. In Part 3, Descartes adopts what he calls a "provisional morality." Why does he do this? How does the adoption of this provisional morality distinguish Descartes's method from other forms of skepticism? Can one properly question one's most basic beliefs while operating on a pre-established set of practical principles? What is the relationship of Descartes's method to morality—does Descartes consider ethics ultimately just as susceptible to definite proof as other bodies of knowledge, such as medicine and geometry?

3. "[T]he soul, by which I am what I am, is entirely distinct from the body and is even easier to know than the body, and even if the body did not exist, the soul could not help being everything it is." This is a crucial claim for Descartes, and one of the ways in which he makes a distinct break from the Aristotelianism of his and earlier ages. What exactly does Descartes take himself to have established here?

4. Some readers are surprised to discover, later in the *Discourse*, that Descartes denies that the soul is "lodged in the human body like a pilot in their ship" but asserts instead that it is "necessary that the soul be joined and united more closely with the body, so that it has, in addition, feelings and appetites similar to ours and thus comprises a true human being." What do you think Descartes means by this? Does it complicate his famous dualism of the mind and body?

5. Consider Descartes's three arguments for God's existence in Part 4. Are these arguments persuasive? Does Descartes's reasoning rely on any hidden premises, or assumptions that might strike modern readers as implausible? Descartes writes that "if we did not know that everything real and true within us comes from a perfect and infinite being, then no matter how clear and distinct our ideas were, we would not have a single reason to assure us that they had the perfection of being true." Must Descartes make this concession? How central is it to his method?

6. Descartes writes, in Part 5, "[f]rom there I went on to speak in particular about the earth: about how, although I had expressly assumed that God had placed no heaviness in the material of which it is composed, all its parts could not help tending precisely to its center." What do you think Descartes means by saying that the planets would naturally form from chaotic matter even if "God had placed no heaviness in the material"?

7. In Part 5, Descartes claims that no machine could fool us into thinking it was a human being. One of his arguments for this is that the machine would be incapable of communicating in the way that humans do. "For one can easily imagine a machine made in such a way that it expresses words ... but one cannot imagine a machine that arranges words in various ways to reply to the sense of everything said in its presence, as the most stupid human beings are capable of doing." This test for thinking interestingly foreshadows more recent ideas in artificial intelligence, such as Alan Turing's "imitation game," in which a person communicates with two interlocutors—one a human and one a computer—and attempts to determine which is which. If a machine were in fact capable of fooling us in the imitation game, such that we would be unable to distinguish it from a human being (which some claim has already happened), would this undermine Descartes's position? Would it show that a complicated machine can in fact think, or that this test for thinking is not a good one? What bearing would it have on Descartes's more general claims about the differences between humans, non-human animals, and machines?

8. Descartes argues, in Part 5, that non-human animals, no matter how intelligent they may seem, don't merely have less reason than humans but in fact have none at all (while, by contrast, even the least intelligent-seeming human being does have reason). Given the role of reason in Descartes's worldview, what are the implications of this

sharp distinction between human and other animals? Is Descartes required, by his own reasoning, to come to this conclusion?

9. Descartes understood his method as one that would lead us to certain beliefs rather than merely probable ones. Yet, many of Descartes's philosophical arguments (for God's existence, for example) are disputed, and many of his scientific conclusions (about the details of how the heart works, and on other topics outside of the *Discourse*) are now known to be false. Is Descartes's defense of his methodology undermined by its apparent failure to provide the kind of certainty that he aspires to?

10. Is it the job of science, in Descartes's view, to explain effects in terms of their causes, or to discover causes from their effects, or somehow both?

11. Many modern scientists treat *fallibility* as an essential methodological principle. That is, they operate on the assumption that it is at least logically possible for any scientific claim to be falsified by future observations. Is this understanding of fallibility compatible with Descartes's approach to knowledge?

12. Part 6 of the *Discourse* contains a lengthy discussion that seems to argue against the benefit of publishing the results of scientific experiments for others to replicate and build upon. "In a word, if there is in the world some work which cannot be properly completed by anyone other than the same person who started it, it's the work I do." What does this suggest about Descartes's view of the cumulative nature of science? Is this a problem for the application of Descartes's method to modern science?

DISCOURS

DE LA METHODE

Pour bien conduire ſa raiſon, & chercher la verité dans les ſciences.

PLUS

LA DIOPTRIQVE.

LES METEORES.

ET

LA GEOMETRIE.

Qui ſont des eſſais de cete METHODE.

A LEYDE

De l'Imprimerie de IAN MAIRE.

cIↄ Iↄ c XXXVII.

Auec Priuilege.

Discourse on the Method

for Reasoning Well and for Seeking Truth in the Sciences

If this discourse seems too long to be read in a single sitting, it can be divided up into six parts. In the first will be found various considerations concerning the sciences; in the second, the principal rules of the method which the author has discovered; in the third, some rules of morality which he has derived by this method; in the fourth, the reasons whereby he proves the existence of God and of the human soul, which are the foundations of his metaphysics; in the fifth part, the order of questions in physics which he has looked into, and particularly the explanation for the movements of the heart and for some other difficulties which are part of medicine, including the difference which exists between our souls and those of animals; and in the last part, some matters he believes necessary for further advances in research into nature, beyond where he has been, along with the reasons that induced him to write.

PART ONE

2 The most widely shared thing in the world is good sense, for everyone thinks they are so well provided with it that even those who are the most difficult to satisfy in everything else do not usually desire to have more good sense than they have. In this matter it is unlikely that everyone is mistaken. But this is rather a testimony to the fact that the power of judging well and distinguishing what is true from what is false, which is really what we call good sense or reason, is naturally equal in all people, and thus the diversity of our opinions does not arise because some people are more reasonable than others, but only because we conduct our thoughts by different routes and do not consider the same things. For it is not enough to have a good mind; the main thing is to apply it well. The greatest minds are capable of the greatest vices as well as the greatest virtues, and those who proceed only very slowly, if they always stay on the right road, are capable of advancing a great deal further than those who rush along and wander away from it.

As for myself, I have never presumed that my mind was anything more perfect than the ordinary mind. I have often even wished that I could have thoughts as quick, an imagination as clear and distinct, or a memory as ample or as actively involved as some other people. And I know of no qualities other than these which serve to perfect the mind. As far as reason, or sense, is concerned, given that it is the only thing which makes us human and distinguishes us from the animals, I like to believe that it is entirely complete in each person, following in this the common opinion of philosophers, who say that differences of more and less occur only between *accidental characteristics* and not between the *forms* or natures of 3 individuals of the same species.

But I will not hesitate to state that I think I have been very fortunate to have found myself since my early years on certain roads which have led me to considerations and maxims out of which I have created a method by which, it seems to me, I have a way of gradually increasing my knowledge, raising it little by little to the highest point which the mediocrity of my mind and the short length of my life will allow it to attain. For I have already harvested such fruit from this method that, even though, in judging myself, I always try to lean towards the side of distrust rather than to that of presumption, and although, when I look with a philosopher's eye on the various actions and enterprises of all people, there are hardly any which do not seem to me vain and useless, I cannot help deriving extreme satisfaction from the progress which I think I have already made in seeking truth and conceiving such hopes for the future that, if among the occupations of people, considered simply as people, there is one which is surely good and important, I venture to think it is the one I have chosen.

However, I may be mistaken and perhaps what I have taken for gold and diamonds is only a little copper and glass. I know how much we are subject to making mistakes in what concerns ourselves and also how much we should be wary of our friends' judgments when they are in our favor. But I will be only too happy to make known in this discourse what roads 4 I have followed and to reveal my life in it, as if in a picture, so that each person can judge it. Learning from current reports the opinions people have of this discourse may be a new way of educating myself, something I will add to those which I habitually use.

Thus, my purpose here is not to teach the method which everyone should follow in order to reason well, but merely to reveal the way in which I have tried to conduct my own reasoning. Those who take it upon themselves to give precepts must consider themselves more skilful than those to whom they give them, and if they fall short in the slightest thing then they are culpable. But since I intend this text only as a history, or, if you prefer, a fable, in which, among some examples which you can imitate, you will also perhaps find several others which you will have reason not to follow, I hope that it will be useful to some people, without harming anyone, and that everyone will find my frankness agreeable.

I was nourished on literature from the time of my childhood, and because I was persuaded that through literature one could acquire a clear and assured understanding of everything useful in life, I had an intense desire to take it up. But as soon as I had completed that entire course of study at the end of which one is usually accepted into the rank of scholars, I changed my opinion completely. For I found myself burdened by so many doubts and errors that it seemed to me I had gained nothing by trying to instruct myself, other than having increasingly discovered my own ignorance. Yet I was in one of the most famous schools in Europe, 5 a place where I thought there must be erudite people, if there were such people anywhere on earth. I had learned everything which the others learned there; still, not being satisfied with the sciences we were taught, I had gone through all the books I could lay my hands on dealing with those sciences[1] which are considered the most curious and rare. In addition, I knew how other people were judging me, and I saw that they did not consider me inferior to my fellow students, although some among

1 The word "science" in Descartes's vocabulary means any formally organized theoretical knowledge. It does not refer merely to the natural sciences.

them were already destined to fill the places of our teachers. And finally, our age seemed to me as flourishing and as fertile in good minds as any preceding age. Hence, I took the liberty of judging all others by myself and of thinking that there was no doctrine in the world of the kind I had previously been led to hope for.

However, I did not cease valuing the exercises which kept people busy in the schools. I knew that the languages one learns there are necessary for an understanding of ancient books; that the gracefulness of fables awakens the intellect; that the memorable actions of history raise the mind and, when read with discretion, help to form one's judgment; that reading all the good books is like having a conversation with the most honorable people of past centuries, who were their authors, even a carefully prepared dialogue in which they reveal to us only the best of their thoughts; that eloquence has incomparable power and beauty; that poetry 6 has a most ravishing delicacy and sweetness; that mathematics has very skillful inventions which can go a long way toward satisfying the curious as well as facilitating all the arts and lessening the work of people; that writings which deal with morals contain many lessons and many exhortations to virtue which are extremely useful; that theology teaches one how to reach heaven; that philosophy provides a way of speaking plausibly on all matters and of making oneself admired by those who are less scholarly; that jurisprudence, medicine, and the other sciences bring honor and riches to those who cultivate them; and finally that it is good to have examined all of them, even the most superstitious and false, in order to know their legitimate value and to guard against being wrong.

But I believed I had already given enough time to languages and even to reading ancient books as well, and to their histories and fables. For speaking with those from other ages is almost the same as traveling. It is good to know something about the customs of various people, so that we can judge our own more sensibly and do not think everything different from our own ways ridiculous and irrational, as those who have seen nothing are accustomed to doing. But when one spends too much time travelling, one finally becomes a stranger in one's own country, and when one is too curious about things which went on in past ages, one usually lives in considerable ignorance about what goes on in this one. In addi- 7 tion, fables make us imagine many impossible events as possible, and even the most faithful histories, if they neither change nor increase the importance of things to make them more worth reading, at the very least almost always omit the most menial and less admirable circumstances, with the result that what remains does not depict the truth. Hence, those who regulate their habits by the examples which they derive from these histories are prone to fall into the extravagances of the knights of our romances and to dream up projects which are beyond their powers.

I placed a great value on eloquence, and I was in love with poetry, but I thought that both were gifts of the mind rather than fruits of study. Those with the most powerful reasoning and who best process their thoughts in order to make them clear and intelligible can always convince us best of what they are proposing, even if they speak only the language of Lower Brittany[2] and have never learned rhetoric. And those with the most pleasant creative talents and who know how to express them with the most adornment and sweetness cannot help being the best poets, even if the art of poetry is unknown to them.

I found mathematics especially delightful because of the certainty and clarity of its reasoning. But I was not yet aware of its true use. Thinking that it was practical only in the mechanical arts, I was astonished that on its foundations, so strong and solid, nothing more imposing had been built. In contrast, I compared the writings of the ancient pagans[3] which deal with morality to really superb and magnificent palaces built on nothing but sand 8 and mud. They raise the virtues to a very great height and make them appear valuable, above everything in the world, but they do not adequately teach us to know them, and often what they call by such a beautiful name is only insensibility or pride or despair or parricide.[4]

I revered our theology and aspired as much as anyone to reach heaven; however, having learned, as something very certain, that the road leading there is no less open to the most ignorant as to the most learned, and that the revealed truths which lead there are beyond our intelligence, I did not dare to submit them to the frailty of my reasoning, and I thought that undertaking to examine them successfully would require some kind of extraordinary heavenly assistance and to be more than human.

I will say nothing of philosophy other than this: once I saw that it has been cultivated for several centuries by the most excellent minds which have ever lived, and that, nonetheless, there is still nothing in it which is not disputed and which is thus not still in doubt, I did not have sufficient presumption to hope to fare better there than the others. Moreover, considering how many different opinions, maintained by learned people, philosophy could have about the same matter—though no more than one

---

2 Descartes is referring to the Breton language, which was more typical among commoners than nobility and may have served as a marker of lower education.

3 In this context, "pagans" refers to non-Christians—thus the ancient Greeks.

4 The word "parricide" may seem odd here, but it refers to acts committed against one's own family in the name of justice (i.e., a love of justice so strong that one is willing to kill members of one's own family who have done wrong). Certain pagan moralists considered such acts particularly virtuous.

could ever be true—I reckoned as virtually false all those which were merely probable.

Then, as for the other sciences, insofar as they borrow their principles 9 from philosophy, I judged that nothing solid could have been built on such insubstantial foundations, and neither the honor nor the profit which they promise were sufficient to convince me to learn them; for, thank God, I did not feel myself in a condition which obliged me to make a profession of science in order to improve my fortune, and, although I did not, in some cynical way, undertake to proclaim my disdain for glory, nonetheless I placed very little value on the fame I could hope to acquire only through false titles. And finally, as for bad doctrines, I thought I already sufficiently understood what they were worth in order not to be taken in by the promises of an alchemist, by the predictions of an astrologer, by the impostures of a magician, or by the artifice or the bragging of any of those who professed to know more than they know.

That is why, as soon as I was old enough to leave the supervision of my professors, I ceased the study of letters, and, resolving to stop seeking any other science except one which could be found inside myself or in the great book of the world, I spent the rest of my youth traveling, looking into courts and armies, associating with people of various humors and conditions, collecting various experiences, testing myself in those encounters which fortune offered me, and everywhere reflecting on the things I came across in such a way as to draw some profit from them. For it seemed to me that I could arrive at considerably more truth in the reasoning that each person makes regarding matters which are important to them and in which 10 events could punish them soon afterwards if they judged badly, than in the reasoning made by a scholar in their study concerning speculations which produce no effect and which are of no consequence to them, except perhaps that from them they can increase their vanity—and all the more so, the further their speculations are from common sense, because they would have had to use that much more wit and artifice in attempting to make them probable. And I always had an extreme desire to learn to distinguish the true from the false, in order to see clearly in my actions and to proceed with confidence in this life.

It is true that while I did nothing but examine the customs of other people, I found hardly anything there to reassure me, and I noticed as much diversity among people as I had earlier noted among the opinions of philosophers. Consequently, the greatest profit which I derived from this was that, upon seeing several things which, although they seem really extravagant and ridiculous to us, were commonly accepted and approved by other great peoples, I learned not to believe too firmly in anything which I had been persuaded to believe merely by example and by custom. Thus, I gradually freed myself of numerous errors which can obfuscate our natural light[5] and make us less capable of listening to reason. But after I had spent a few years studying in this way in the book of the world and attempting to acquire some experience, one day I resolved to study myself as well and to use all the powers of my mind to select paths which I should follow, a task which brought me considerably 11 more success, it seems to me, than if I had never left my own country or my books.

5 *natural light*: our innate faculty of rational understanding.

PART TWO

I was then in Germany, summoned by the ongoing wars there.[6] As I was returning to the army from the emperor's coronation,[7] the onset of winter stopped me in a place where—not finding any conversation to divert me and, in addition, by good fortune, not having any cares or passions to trouble me—I spent the entire day closed up alone in a stove-heated room, where I had complete leisure to talk to myself about my thoughts. Among these thoughts, one of the first I noticed was how often there is not as much perfection in works created from several pieces and made by the hands of various masters as there is in those which one person has worked on alone. Thus, we see that the buildings which a single architect has undertaken and completed are usually more beautiful and better ordered than those which several people have tried to refurbish by making use of old walls built for other purposes. That is why those ancient cities which were only small villages at the start and became large towns over time are usually so badly laid out compared to the regular places which an engineer has designed freely on level ground. Even though, considering the buildings in each of them separately, we often find as much beauty in the former town as in the latter, or more, nonetheless, looking at them as they are arranged—here a large one, there a small one—and the way they make the streets crooked 12 and unequal, we would say that chance rather than the will of some people using their reason designed them this way. And if one considers that nonetheless there have always been certain officials charged with seeing that private buildings serve as a public ornament, one will readily see that it is difficult to achieve really fine things by working only with other people's pieces. Thus, I imagined to myself that people who were semi-savages in

---

6 In 1618 Descartes, who was Catholic, voluntarily joined the Protestant army of Maurice of Nassau, who was active in organizing the forces of the Dutch Republic in its fight against Spain. In 1619, however, he left the Dutch army and traveled to join the army of Maximilian of Bavaria in Germany, which was part of the "Catholic League" fighting Protestant forces in Bohemia and Upper Austria. This conflict was part of the Thirty Years' War (1618–48), one of the most devastating wars in European history, which ultimately resulted in the deaths of more than eight million people.

7 This was the coronation of the (zealously anti-Protestant) Ferdinand II as Holy Roman Emperor, which took place in Frankfurt in late summer 1619.

earlier times and who became civilized only little by little and created their laws only as they were compelled to by the extent to which crimes and quarrels bothered them, would not be so well regulated as those who, from the moment they first assembled, followed the constitution of some prudent legislator. It is indeed certain that the state of the true religion, whose laws God alone created, must be incomparably better ordered than all the others. And, in terms of human affairs, I believe that if Sparta was in earlier times very prosperous, that was not on account of the goodness of each of its laws in particular, seeing that several were very strange and even contrary to good morals, but because they were devised by only a single man and thus aimed at the same end.[8] Similarly, I thought that the sciences contained in books—at least those whose reasons are only probable and without any proofs, since they were composed and fashioned little by little out of the opinions of several different people—therefore did not approach the truth as much as the simple reasonings which a person of good sense 13 can make quite naturally concerning matters of their own experience. In the same way, I thought that because we were all children before we were adults and because it was necessary for us to be governed for a long time by our appetites and our supervisors, who were often at odds with each other, with neither of them perhaps advising us always for the best, it is almost impossible for our judgments to be as pure and solid as they would have been had we had the total use of our reason from the moment of our birth and never been led by anything but our reason.

It is true that we see little point in demolishing all the houses of a city for the sole purpose of rebuilding them in another way and thus making the streets more beautiful. However, we do see many people demolish their houses in order to rebuild them, and, indeed, sometimes they are compelled to do so, when the houses are in danger of collapsing on their own and when their foundations are unstable. This example persuaded me that there would probably be little point in a particular person drawing up a design to reform a state by changing everything starting with the foundations, and overturning it in order to put it up again, or even in reforming the body of sciences or the order established in the schools for teaching the sciences. But so far as all the opinions which I had received up to that point and which I believed credible were concerned, I convinced myself that the best possible thing for me to do was to undertake to remove them once and for all, so that afterwards I could replace them either by other, better ones or perhaps by the same ones, once I had adjusted them to a 14

8 Sparta was an ancient Greek city-state with social structures and laws designed to emphasize military proficiency. Its laws were supposedly created by a single person, Lycurgus, though there is some uncertainty as to whether he actually existed.

reasonable standard. And I firmly believed that by this means I would be successful in conducting my life much better than if I built only on old foundations and relied only on principles which I had been persuaded to accept in my youth, without ever having examined whether they were true. For, although I recognized various problems with this approach, these were not without remedy and could not compare to those which occur in reforming the least matters concerning the public. It is too difficult to re-erect those large bodies if they are thrown down or even to keep them once they are weakened, and their collapse cannot be anything but very drastic. Then, as far as the imperfections of large public bodies are concerned, if they have any (and the variety among such bodies alone is enough to assure us that many have some), habit has no doubt considerably softened them and has even managed to avoid some problems or corrected a number of them insensibly, which people's caution could not have managed so well. And finally, the imperfections are almost always easier to bear than changing them would be, in the same way that the major roads which wind among the mountains gradually become so smooth and convenient from being used that it is much better to follow them than to attempt to go more directly by climbing up over the rocks and going down to the very bottom of the precipices.

That is why I cannot approve at all of those muddled and worried temperaments who, being summoned neither by birth nor fortune to the
15 management of public business, never stop proposing some idea for a new reform in it. If I thought that there was the slightest thing in this text which would enable someone to suspect me of this foolishness, I would be very reluctant to allow it to be published. My intention has never been to do more than try to reform my own thoughts and to build on a foundation which is entirely my own. And if my work has pleased me sufficiently to make me show you the model of it here, that is not because I wish to advise anyone to imitate it. Those to whom God has given more of His grace will perhaps have loftier intentions, but I fear that this work may already be too bold for many people. The single resolution to strip away all the opinions which one has previously absorbed into one's beliefs is not an example which everyone should follow. Most of the world is made up of two kinds of minds for whom such a resolution is not suitable. First, there are those who, believing themselves more clever than they are, cannot stop making hasty judgments, without having enough patience to conduct their thoughts in an orderly manner, with the result that, once they have taken the liberty of doubting the principles they have received and of leaving the common road, they will never be able to keep to the path which they must take in order to proceed more directly, and will remain lost all their lives. Then, there are those who, having sufficient reason or modesty to judge that they are less capable of differentiating truth from falsehood than others from whom they

can be instructed, must content themselves with following the opinions of these others rather than searching for better opinions on their own.

As for me, I would have undoubtedly been among those in this latter group if I had only had a single master or if I had known nothing at all about the differences which have always existed among the opinions of the most highly educated people. But I learned from my college days on that one cannot imagine anything so strange and so incredible that it has not been said by some philosopher and, later, in my travels, I recognized that all those who have views very different from our own are not therefore barbarians or savages, but that many of them use as much reason as we do, or more. I also considered how much the same person, with the same mind, raised from infancy on among the French or the Germans, would become different from what they would have been had they always lived among the Chinese or the cannibals, and how, even in our style of dress the same thing which pleased us ten years ago and which will perhaps please us again ten years from today, now seems to us extravagant and ridiculous. Thus we are clearly persuaded more by custom and example than by any certain knowledge. Nonetheless, a plurality of voices is not a proof worth anything for truths which are somewhat difficult to discover, because it is far more probable that one person by themselves would have found them than an entire people. Since I could not select anyone whose opinions it seemed to me one should prefer to those of other people, I found myself, so to speak, compelled to guide myself on my own. 16

But like a person who proceeds alone and in the shadows, I resolved to go so slowly and to use so much circumspection in all matters that, if I only advanced a very short distance, at least I would take good care not to fall. I did not even wish to begin completely rejecting any of the opinions which could have slipped into my beliefs previously without being introduced by reason, until I had spent enough time drawing up a plan for the work I was undertaking and seeking out the true method for arriving at an understanding of everything my mind was capable of knowing. 17

When I was younger, among the branches of philosophy, I had studied a little logic and, among the subjects of mathematics, geometrical analysis, and algebra—three arts or sciences which looked as if they ought to contribute something to my project. But in examining them, I was cautious because, so far as logic is concerned, its syllogisms[9] and most of its other instructions serve to explain to others what one already knows or even, as

9 Syllogisms are the forms of deductive reasoning associated with Aristotle. These each had two premises and one conclusion, e.g., "All men are mortal; Socrates is a man; therefore Socrates is mortal." In a sense, these do not lead to new knowledge: if you already knew the premises, wouldn't you also know the conclusion?

in the art of Llull,[10] to speak without judgment of things about which one is ignorant, rather than to learn what they are. Although philosophy does, in fact, contain many really true and excellent precepts, mixed in with them there are always so many injurious or superfluous ones that it is almost as difficult to separate them as to draw a Diana or a Minerva out of a block of marble which has not yet been carved.[11] Then, so far as the analysis of the ancients and the algebra of the moderns are concerned, aside from the fact that they deal only with really abstract matters that have no apparent use, the former is always so concentrated on considering numbers that it 18 cannot exercise the understanding without considerably tiring the imagination, and the latter is so subject to certain rules and symbols that it has been turned into a confused and obscure art which clutters up the mind rather than a science which cultivates it. Those were the reasons why I thought I had to look for some other method which included the advantages of these three subjects but was free of their defects. And since a multitude of laws often provides excuses for vices, so that a state is much better ruled when it has only very few laws which are very strictly observed, I thought that, instead of that large number of rules which comprise logic, I would have enough with the four following rules, provided that I maintained a strong and constant resolution never to fail to observe them, not even once.

The first rule was that I would not accept anything as true which I did not clearly know to be true. That is to say, I would carefully avoid being over-hasty or prejudiced, and I would understand nothing by my judgments beyond what presented itself so clearly and distinctly to my mind that I had no occasion to doubt it.[12]

The second was to divide each difficulty which I examined into as many parts as possible and necessary to resolve it better.

---

10 Ramon Llull (c. 1232 – c. 1315) was a Majorcan philosopher and mathematician who in multiple works developed a system he called "the art," which was designed to demonstrate the truth of Christianity using logic and visual aids. Llull also developed the first Condorcet method of election.

11 Diana is the ancient Roman goddess of the hunt, Minerva the goddess of wisdom and warfare. Each has been the subject of numerous sculptures.

12 Descartes's *Principia Philosophiae*, 1:45–46, discusses (in Latin) his use of these terms: "I call an idea clear [*claram*] when it is present and manifest to a mind focusing on it, just as we say we perceive something clearly when it is present to the observing eye, and stimulates it sufficiently strongly and fully. I call an idea distinct [*distinctam*] which, while it is clear, is separated and marked off from everything else in such a way that it consists of absolutely nothing which is not clear." Descartes may have in mind the clarity and distinctiveness of geometrical propositions.

The third was to conduct my thoughts in an orderly way, beginning with the simplest objects, the ones easiest to know, so that little by little I could gradually climb right up to the knowledge of the most complex, even assuming an order among those things which do not naturally come 19 one after the other.

And the last was to make my calculations throughout so complete and my reviews so general that I would be confident of not omitting anything.

Those long chains of reasons, all simple and easy, which geometers have habitually used to reach their most difficult proofs, gave me the opportunity to imagine to myself that everything which could fall under human knowledge would follow in the same way and that, provided only that one refused to accept anything as true which was not, and that one always kept to the order necessary to deduce one thing from another, there could not be anything so distant that one could not finally reach it, nor so hidden that one could not discover it. And I did not have much trouble determining the issues which I had to deal with first, for I already knew that I had to begin with the simplest things, the ones easiest to know. When I thought about how, among all those who had thus far sought truth in the sciences, only the mathematicians had been able to find some proofs—that is to say, some certain and evident reasons—I had no doubt that I should start with the same things which they had examined, although I did not hope for any practical results other than that they would accustom my mind to revelling in the truth and not being satisfied with false reasons. But for all that, I did not plan on trying to learn all the particular sciences which people commonly call mathematical,[13] as I saw that, even though 20 their objects were different, they were alike in that they all agreed they should consider nothing except the various relationships or proportions among the objects of study found there. Thus I thought it best to examine only these proportions in general, considering them only in the objects that would most readily help to provide me with knowledge of them, but without in this way restricting them whatsoever, so that they could be all the better applied later to every other object for which they might be suitable. Then, because I observed that, in order to understand these things, I would sometimes need to consider each one in particular and sometimes only to remember them or to understand several of them together, I thought that to consider them better separately, I ought to assume that they were like lines, because I knew of nothing simpler, nothing which I could more distinctly represent to my imagination and my senses. But in order to remember them or to understand several of them together, I had to explain them by some formulas as short as possible and in so doing would borrow all the

13 Such as optics, astronomy, mechanics, and music (all of which Descartes did in fact study).

best elements of analytic geometry and algebra and would correct all the defects of one using the other.[14]

As a matter of fact, I venture to say that the precise observation of these few precepts which I had selected made it so easy for me to disentangle all the questions which these two sciences cover, that in the two or three months that I used them to examine these questions, having begun with the simplest and the most general and letting each truth I found serve as a rule 21 which I could use afterwards to find others, not only did I resolve several problems which I had previously judged very difficult, but it also seemed to me towards the end that I could determine, even with those questions where I was ignorant, the way to resolve them and the extent to which such resolution was possible. In saying this, perhaps I will not appear too vain if you consider that, since there is only one truth for each thing, whoever finds it knows as much as one can know about it and that, for example, a child instructed in arithmetic, having made an addition following the rules, can be confident of having found, so far as the sum they are examining is concerned, everything that the human mind can find out. For the method which teaches one to follow the true order, and to count exactly all the relevant details in what one is looking for, contains everything which gives certainty to the rules of arithmetic.

But what pleased me the most with this method was that with it I was confident of using all my reason, if not perfectly, at least as well as was in my power. In addition, I felt, as I applied it, that my mind was accustoming itself gradually to thinking more clearly and distinctly about its objects, and because I had not restricted this method to one matter in particular, I was hopeful that I could apply it just as usefully to difficulties in the other sciences as I had applied it to those in algebra. But for all that, I did not venture to try to immediately examine all those scientific problems which presented themselves, for that would have been contrary to the order which my method prescribed. I noticed that the principles of science all had to 22 be borrowed from philosophy, a subject in which I no longer found anything certain. So I thought that, before anything else, I should attempt to establish such principles there and that, since this was the most important matter in the world, where one had to be most fearful of overhasty and biased judgments, I would not try to get through it until I had reached an age considerably more mature than I was then at twenty-three and until I had spent a lot more time preparing myself, weeding out of my mind all the bad opinions which I had accepted before that time, as well as collecting many experiences so that later they could be the subject matter of my reasoning, always practicing the method which I had set for myself in order to keep on improving myself in these matters.

14 Descartes here is referring to his discoveries in analytic geometry, in which algebraic equations are used to represent geometric relations.

PART THREE

Finally, before one starts to rebuild the lodgings where one lives, it is not enough to knock them down and provide for materials and architects or to work on the architecture oneself, having, in addition, carefully drawn up a design. One must also provide oneself with some other place where one can lodge comfortably while one works on the building. Thus, in order not to be irresolute in my actions while my reason obliged me to be so in my judgments, and in order not to prevent myself living from then on as happily as I could, I drew up for myself a provisional morality, consisting of only three or four maxims, which I wish to share with you.

The first was to obey the laws and the customs of my country, constant- 23 ly holding to the religion which God gave me the grace to be instructed in since childhood and governing myself in all other things in accordance with the most moderate opinions, the ones furthest removed from excess, which were commonly accepted and practiced by the most sensible of those people among whom I would be living. Since, from that point on, I began to estimate my own views as worthless, because I wished to subject them all to examination, I was confident that I could not do better than to follow those of the most sensible people. And even though there might perhaps be people just as sensible among the Persians or the Chinese as among us, it seemed to me that the most practical thing would be for me to guide myself by those among whom I had to live and that, in order to understand their real opinions, it would be better for me to pay attention to what they practiced rather than to what they said, not only because, given the corruption of our morals, few people are willing to state everything they believe, but also because several are themselves ignorant of what they believe. For the act of thinking by which one believes in something is different from the act of thinking by which one understands that one believes it, and one of these separate acts frequently appears without the other. Moreover, among opinions equally well received, I chose only the most moderate ones, as much because such opinions are always the most convenient to practice and probably the best—for all excess is usually bad—as because they would also not divert me as much from the true road, if I made a mistake, as if I had chosen one of the extremes when it was the other one which I should have followed. And I especially included among 24 what was excessive all those promises by which one reduces one's liberty. Not that I disapprove of laws which, in an attempt to remedy the fickleness of feeble minds, allow people with a good plan or even an indifferent arrangement for security in business to make vows or contracts obliging them to maintain their provisions. But because I did not see anything in the world which remained always in the same condition and, in my particular case, because I promised myself that I would increasingly perfect my judgments and not make them worse, I would have thought I was committing a great error against good sense if, because I then approved of something, I obliged myself to continue to take it as something good later on, when it had perhaps ceased to be so or when I had ceased to value it as something good.

My second maxim was to be as constant and as resolute in my actions as I could, and to follow the most doubtful opinions, once I had settled on them for myself, with no less constancy than if they had been very sure, imitating in this matter travelers who, finding themselves lost in some forest, should not wander around, shifting direction this way and that; even less should they stop in one place; they should move on always as straight as they can in the same direction and not change their course for inadequate reasons, even though at the beginning it was perhaps only chance which led to their choice of direction. For in this way, if they do 25 not come out exactly where they want to, they will at least end up arriving somewhere where they will probably be better off than in the middle of a forest. And because the actions of life often brook no delay, it is certainly very true that, when it is not in our power to determine the truest opinions, we ought to follow the most probable ones, and even when we see no difference in probability among this or that group of truths, we must nonetheless decide on some for ourselves and then consider them no longer as something doubtful with regard to the practical matter at hand, but as manifestly true and very certain, because the reason which made us choose them has these qualities. This method was able from then on to relieve me of all the regrets and remorse which usually upset the consciences of those weak and wavering minds who allow themselves to work inconsistently with things which they accept as good but which they later judge to be bad.

My third maxim was to try always to overcome myself rather than fortune and to change my desires rather than the order of the world, and generally to get in the habit of believing that there is nothing which is entirely within our power except our thoughts, so that after we have done our best regarding those things which lie outside of us, everything which our attempt fails to deal with is, so far as we are concerned, absolutely impossible.[15] That alone seemed to me to be sufficient to prevent me from

15 This, and the previous maxim, strongly resemble the Roman moral philosophy of Stoicism. Here, for example, are the opening lines of the *Enchiridion* of Epictetus, written in about 125 CE: "Of all existing things some are in our

desiring anything in the future which I might not achieve, and thus to make me happy. For since our will has a natural tendency to desire only 26 things which our understanding represents as in some way possible, it is certain that if we think about all the good things which are outside of us as equally distant from our power, we would no more regret missing those whose loss appears due to our birth, when we are deprived by no fault of our own, than we would regret not possessing the kingdoms of China or Mexico. By making, as the saying goes, a virtue of necessity, we would not desire health when we are sick or freedom when we are in prison, any more than we now desire to have either a body made of some material as incorruptible as diamonds, or wings to fly like the birds. But I admit that long discipline and frequently repeated meditation are needed in order to accustom oneself to looking at everything from this point of view. And I believe that this is the principal secret of those philosophers who have been able in earlier times to escape from the demands of empire and fortune and who, despite pains and poverty, could rival their gods in happiness. For, constantly busy thinking about the limits prescribed for them by nature, they persuaded themselves so perfectly that nothing was in their power except their thoughts, that that alone was enough to prevent them from having any affection for other things; and they acquired such an absolute control over their thoughts that they found in that process reason to think themselves more rich and more powerful and more free and more content than any other person, who, because they did not possess this philosophy, never had the same control over everything they desired, no matter how 27 favored they might be by nature and fortune.

---

power, and others are not in our power. In our power are thought, impulse, will to get and will to avoid, and, in a word, everything which is our own doing. Things not in our power include the body, property, reputation, office, and, in a word, everything which is not our own doing. Things in our power are by nature free, unhindered, untrammelled; things not in our power are weak, servile, subject to hindrance, dependent on others. Remember then that if you imagine that what is naturally slavish is free, and what is naturally another's is your own, you will be hampered, you will mourn, you will be put to confusion, you will blame gods and men; but if you think that only your own belongs to you, and that what is another's is indeed another's, no one will ever put compulsion or hindrance on you, you will blame none, you will accuse none, you will do nothing against your will, no one will harm you, you will have no enemy, for no harm can touch you" (as translated by P.E. Matheson, 1916). Neo-Stoicism was of great interest in France around this time of upheaval and religious warfare; for instance, Justus Lipsius's influential *Introduction to Stoic Philosophy* was published in 1604.

Finally, to conclude these moral precepts, I advised myself to review the various occupations which people have in this life, in an attempt to choose the best; and, without wanting to say anything about the others, I thought that I could do no better than to continue in the very occupation I was engaged in, that is, using all my life to cultivate my reason and to progress as far as I could in a knowledge of the truth, following the method which I had prescribed for myself. I experienced such extreme contentment once I started using this method that I did not believe one could find anything more sweet and innocent in this life. Since every day I discovered through this method some truths which seemed to me quite important and commonly unknown to other people, the satisfaction I got from it so filled my mind that nothing else affected me. Moreover, the three maxims mentioned above were founded only for the plan I had to continue my self-instruction. For, since God has given each one of us some light to distinguish truth from falsehood, I would not have thought I could remain content with other people's opinions for one moment if I had not set out to use my own judgment to examine them when the time was right, and I would not have known how to free myself from scruples in following these opinions if I had not hoped that I would, in the process, not lose any 28 opportunity to find better ones in cases where they existed. Finally I would not have known how to limit my desires, nor how to rest content, if I had not followed a road by which I believed I could be confident of acquiring all the knowledge I was capable of. I thought by the same means I could acquire all the true benefits I was capable of obtaining, all the more so since our will tends to follow or to fly away from only those things which our understanding has represented to it as good or bad. Thus, in order to act well it is sufficient to judge well, and to judge as well as one can is sufficient to enable one to do one's best, that is, to acquire all the virtues, along with all the other benefits which one can get. When one is certain that that is the case, one could not fail to be content.

After assuring myself of these maxims in this manner and storing them away along with the truths of the faith which have always been first in my beliefs, I judged that, so far as all the rest of my opinions were concerned, I could freely set about dispensing with them. Since I hoped to be able to arrive at my goal more easily by speaking with other people than by remaining any longer closed up in the room with the stove where I had had all these thoughts, before the winter was over and done with I set about traveling again. And in all the nine years following I did nothing else but roam here and there in the world, trying to be a spectator rather than an actor in all the comedies[16] playing themselves out there. By reflecting on

16 In this older usage, a comedy is a dramatic performance (not necessarily one which is humorous), often involving the resolution of some sort of everyday conflict.

each matter, in particular on what there was which could render it suspect or give us an opportunity to make mistakes, I rooted out from my mind all the errors which could have previously slid into it. Not that in the process 29 I copied the skeptics, who doubt only for the sake of doubting and pretend that they are always irresolute. On the contrary, my entire plan tended only to make me confident about discarding the shifting ground and the sand in order to find the rock or the sedimentary clay. This gave me considerable success, it seems to me, inasmuch as in my attempts to discover the falsity or the uncertainty of the propositions I examined, not by weak conjectures, but by clear and confident reasoning, I came across nothing so doubtful that I did not always draw some fairly certain conclusion from it, even if that conclusion was that it contained nothing certain. Just as when we tear down an old lodging, we usually keep the remnants to use in building a new structure, so, as I destroyed all those opinions of mine which I judged poorly grounded, I made various observations and acquired several experiences which were of use to me later in establishing more certain ones. In addition, I continued to practice the method which I had set for myself. For, aside from the fact that I took care, in general, to conduct all my thinking according to the rules, from time to time I set aside a few hours which I used to apply the method to mathematical difficulties in particular, or even to some other difficulties as well, ones which I could frame in a manner somewhat similar to those in mathematics by stripping from them all the principles of the other sciences which I did not find sufficiently strong, as you will see I have done in several which are explained in this volume.[17] Thus, without living in a way apparently different from those who have 30 nothing else to do but spend a sweet and innocent life studying how to separate pleasures from vices and enjoying their leisure by making use of all honorable entertainments without getting bored, I did not fail to follow my plans and to benefit from the knowledge of the truth, perhaps more so than if I had only read books or associated with scholars.

However, nine years passed by before I had yet taken any stand concerning the difficulties which are usually debated among the scholars.[18] Nor had I begun to seek the foundations of any philosophy more reliable than common philosophy. The example of several excellent minds who had earlier had the same idea but who, it seemed to me, had not succeeded, made me imagine such great difficulties that I would perhaps not have ventured to undertake it so quickly, had I not seen that some people had already spread the rumor that I had concluded my work. I don't know how they arrived at this opinion. And if I contributed something to it by my conversations, that must have been by confessing where I was ignorant

17 In the same book as this *Discourse*, Descartes included sections on optics, geometry, and meteorology (the study of atmospheric phenomena).

18 The "nine years" Descartes refers to are 1619 to 1628.

more ingenuously than those who have studied little are accustomed to doing, and perhaps also by making known the reasons I had for doubting many things which other people considered certain, rather than by boasting about any doctrine. But having a heart sufficiently good not to wish people to take me for someone other than the man I am, I thought it necessary to attempt by every means to make myself worthy of the reputation which 31 people ascribed to me. For exactly eight years this desire made me resolve to distance myself from all those places where there might be people I know and to retire here, in a country where the long duration of the war has established such order that the armies which maintain it appear to serve only to enable the people to enjoy the fruits of peace with even more security and where, among the crowd of a great and very active people, who are more careful about their own affairs than curious about those of other people, and lacking no conveniences found in the most frequently visited towns, I was able to live retired in solitude, just as if I were in the most isolated deserts.[19]

19 Descartes is here referring to his time in the Dutch Republic. The war he is referring to is that between the United Provinces of the Netherlands and Spain, their former ruler, which lasted from 1568 until 1648.

PART FOUR

I don't know if I should share with you the first meditations which I made there, for they are so metaphysical and so out of the ordinary that they will perhaps not be to everyone's taste. However, in order that people may judge if the foundations which I set are sufficiently strong, I find myself in some way compelled to speak of them. For a long time previously I had noticed that where morals are concerned it is sometimes necessary to follow opinions which one knows are extremely uncertain as if they were indubitable, as mentioned above. But since at that time I wanted only to carry out research into the truth, I thought I must do the opposite and reject as absolutely false everything about which I could imagine the least doubt, in order to see if anything totally indubitable would remain after that in my belief. Thus, because our senses deceive us sometimes, I was willing 32 to assume that there was nothing which existed the way our senses present it to us. And because there are people who make mistakes in reasoning, even concerning the most simple matters of geometry, and who create paralogisms,[20] and because I judged that I was subject to error just as much as anyone else, I rejected as false all the reasons which I had taken earlier as proofs. Finally, given that all the same thoughts we have when awake can also come to us when we are asleep, without there being truth in any of them at the time, I determined to pretend that everything which had ever entered my mind was no more true than the illusions of my dreams. But immediately afterwards I noticed that, while I wished in this way to think everything was false, it was necessary that I—who was doing the thinking—had to be something. Noticing that this truth—*I think; therefore, I am*—was so firm and so sure that all the most extravagant assumptions of the skeptics would not be able to weaken it, I judged that I could accept it without scruple as the first principle of the philosophy I was looking for.[21]

20 A paralogism is a fallacious argument—typically, one in which the arguer does not realize the error in their reasoning.

21 In a later work, *Principles of Philosophy*, this statement, "I think; therefore, I am" (*je pense, donc je suis*) becomes the famous Latin sentence *Cogito ergo sum*. As the subsequent lines in the discussion above indicate, this claim might be more properly translated "I am thinking; therefore, I am," since the certainty remains only during the process of thinking.

Then I attentively examined what I was, and saw that I could pretend that I had no body and that the world and the place where I was did not exist, but in spite of this, I could not pretend that I did not exist. By contrast, in the very act of thinking about doubting the truth of other things, it very 33 clearly and certainly followed that I existed; whereas, if I had only stopped thinking, even though all the other things which I had ever imagined were real, I would have no reason to believe that I existed. From that I recognized that I was a substance whose essence or nature is only thinking, a substance which has no need of any location and does not depend on any material thing, so that this "I," that is to say, the soul, by which I am what I am, is entirely distinct from the body and is even easier to know than the body, and even if the body did not exist, the soul could not help being everything it is.

After that, I considered in general what is necessary for a proposition to be true and certain, for since I had just found one idea which I knew to be true and certain, I thought that I should also understand what this certitude consisted of. And having noticed that in the sentence "I think; therefore, I am" there is nothing at all to assure me that I am speaking the truth, other than that I see very clearly that in order to think it is necessary to exist, I judged that I could take as a general rule that the things which we conceive very clearly and very distinctly are all true. But that left the single difficulty of properly noticing which things are the ones we conceive distinctly.

After that, I reflected on the fact that I had doubts and that, as a result, my being was not completely perfect, for I saw clearly that it was a greater perfection to know than to doubt. I realized that I should seek out where I had learned to think of something more perfect than I was, and I concluded 34 that obviously this must be something with a nature which was, in effect, more perfect. As for the thoughts which I had of several other things outside of myself, like the sky, the earth, light, heat, and a thousand others, I was not worried about knowing where they came from, because I did not notice anything in them which seemed to me to make them superior to myself. Thus, I was able to think that, if they were true, that was because of their dependence on my nature, in so far as it had some perfection and, if they were not true, I held them from nothing, that is to say, that they were in me because I had some defect. But that could not be the same with the idea of a being more perfect than mine. For to hold that idea from nothing would be manifestly impossible. And because it is no less unacceptable that something more perfect should be a consequence of and dependent on something less perfect than that something should come from nothing, I could not derive this idea from myself. Thus, I concluded that the idea had been put in me by a nature which was truly more perfect than I was, even one which contained in itself all the perfections about which I could have some idea—that is to say, to explain myself in a single word, a nature which was God. To this I added the fact that, since I know about some perfections which I do not have, I was not the only being which existed (here I will freely use, if you will permit me, the language of the schools[22]), but it must of necessity be the case that there was some other more perfect being, on whom I depended and from whom I had acquired all that I had. For if I had been alone and independent of everything else, so that I derived from myself all perfection, no matter how small, of the 35 perfect being, I would have been able to have from myself, for the same reason, all the additional perfections which I knew I lacked, and thus be myself infinite, eternal, immutable, all knowing, all powerful, and finally have all the perfections which I could observe as present in God. For, following the reasoning which I have just made, to know the nature of God, to the extent that my reasoning is able to do that, I only had to think about all the things of which I found some idea within me and consider whether it was a sign of perfection to possess them or not. And I was confident that none of those ideas which indicated some imperfection were in God, but that all the others were there, since I perceived that doubt, inconstancy, sadness, and similar things could not be in God, given the fact that I myself would have been very pleased to be free of them. Then, in addition, I had ideas about several sensible[23] and corporeal things. For although I supposed that I was asleep and that everything which I saw or imagined was false, nonetheless I could not deny that the ideas had truly been in my thoughts. But because I had already recognized in myself very clearly that intelligent nature is distinct from corporeal nature, when I considered that all composite natures indicate dependency and that dependency is manifestly a defect, I judged from this that God's perfection could not consist of being composed of these two natures, and that thus He was not, but that if there were some bodies in the world or even some intelligences or other natures which were not completely perfect, their being had to depend on 36 God's power, in such a way as they could not subsist for a single moment without Him.

After that I wanted to look for other truths, and I proposed to myself the subject matter of geometricians, which I understood as a continuous body or a space extended indefinitely in length, width, and height or depth, divisible into parts, which could have various figures and sizes and be moved or transposed in all sorts of ways, for the geometricians assume all that in their subject matter. I glanced through some of their simplest proofs, and having observed that this grand certainty which all the world attributes to them is founded only on the fact that they plan these proofs

22 *the language of the schools*: The terminology of scholasticism, the intellectual framework of Descartes's day.

23 *sensible*: able to be sensed—perceivable.

clearly, following the rule which I have so often stated, I noticed also that there was nothing at all in their proofs to assure me of the existence of their objects. So, for example, I saw very well that, if we assume a triangle, it must be the case that its three angles are equal to two right angles; but, in spite of that, I did not see anything to assure me that there is a triangle in the world. By contrast, once I returned to examining the idea which I had of a perfect being, I found that its being contains the idea of existence in the same way as the fact that the three angles of a triangle are equal to two right angles is contained in the idea of a triangle, or that in a sphere all the parts are equidistant from the center, and, as a result, it is at least as certain that God—this perfect being—is or exists, as any geometric proof can be.

37 But the reason many people persuade themselves that there are difficulties in understanding this and even knowing what their soul is, is that they never raise their minds above matters of sense experience and are so accustomed not to consider anything except by imagining it, which is a way of thinking in particular of material things, so that everything which is not imaginable seems to them unintelligible. This point is obvious enough in the fact that even the philosophers in the schools maintain the axiom that there is nothing in the understanding which has not first of all been in the senses.[24] But it is certain that the ideas of God and the soul have never been present in sense experience. It seems to me that those who wish to use their imagination to understand these things are acting just as if they want to use their eyes to hear sounds or smell odors, except that there is still this difference: the sense of sight provides us no less assurance of the truth of what it sees than do the sense of smell or hearing; whereas, neither our imagination nor our senses can assure us of anything unless our understanding intercedes.

Finally, if there are still some people who are insufficiently persuaded of the existence of God and their soul by the reasons I have provided, I would like them to know that everything else which they perhaps are more confident about in their thinking, such as having a body and knowing that there are stars and an earth, and things like that, is less certain than God's existence. For although one has a moral certainty[25] about 38 these things, something which makes doubting them appear at least extravagant, nonetheless, unless one is an unreasonable being, when a question of metaphysical certainty is involved, one cannot deny that there is insufficient material here to make one completely confident, for we

---

24 This was an important principle of scholastic philosophy, taken from Aristotelian science. (It was sometimes known as the "Peripatetic Axiom," after the Peripatetic school of philosophy founded by Aristotle.)

25 *moral certainty*: a high degree of probability, enough to rationally act on, but falling short of absolute certainty.

notice that one can imagine in the same way while sleeping that one has another body and that one sees other stars and another earth, without such things existing. For what is the source of our knowledge that the thoughts which come while dreaming are more likely false than are other thoughts, seeing that often they are no less lively and distinct? And if the best minds study this matter as much as they please, I do not think that they will be able to give any reason which will be sufficient to remove this doubt unless they presuppose the existence of God. First of all, the very principle which I have so often taken as a rule—only to recognize as true all those things which we conceive very clearly and very distinctly—is guaranteed only because of the fact that God is or exists, that He is a perfect being, and that everything which is in us comes from Him. From that it follows that our ideas or notions, being real things which come from God, to the extent that they are clear and distinct, cannot be anything but true. Consequently, if we often enough have some ideas or notions which contain something false, they can only be those which contain some confusion and obscurity, because in this they participate in nothing, that is to say, they are so confused in us only because we are not completely perfect. And it is evident that it is no less repugnant that falsity or imperfection in itself should come from God than that truth or perfection should come from nothingness. But if we did not know that everything real and true within us comes from a perfect and infinite being, then no matter how clear and distinct our ideas were, we would not have a single reason to assure us that they had the perfection of being true. 39

Now, after the knowledge of God and the soul in this way has made us certain of this rule, it is really easy to see that the dreams which we imagine while asleep should not, in any way, make us doubt the truth of the thoughts we have while awake. For if it happened, even while we were sleeping, that we had some really distinct idea, as, for example, in the case of a geometer inventing some new proof, the fact that we were asleep would not prevent it from being true. And as for error, it doesn't matter that the most common dreams we have, which consist of representing to us various objects in the same way as our external senses do, can give us occasion to challenge the truth of such ideas, because these ideas can also mislead us often enough without our being asleep, as, for example, when those suffering from jaundice see all objects as yellow, or when the stars or other bodies at a great distance appear to us much smaller than they are. For, finally, whether we are awake or asleep, we should never allow ourselves to be persuaded except by the evidence of our reason. And people should note that I say of our reason, and not of our imagination or of our senses, since even though we see the sun very clearly, we should not for that reason judge that it is only the size which we see it, and we can easily imagine distinctly the head of a lion mounted on the body of a goat, with- 40 out having to conclude, because of that, there is a chimera[26] in the world: for reason does not dictate to us that what we see or imagine in this way is true, but it does dictate to us that all our ideas or notions must have some foundation in truth, for it would not be possible that God, who is completely perfect and totally truthful, put them in us without that. Because our reasoning is never so evident or complete while we sleep as while we are awake, although sometimes our imaginations when asleep are as vital or explicit or more so, reason also dictates to us that our ideas cannot all be true, because we are not completely perfect; those ideas which contain the truth must without exception be those we experience while awake rather than those we have while asleep.

---

26 In Greek mythology, a female fire-breathing monster with a lion's head, a goat's body, and a serpent's tail.

PART FIVE

I would be very pleased to continue and make you see here all the chain of other truths which I deduced from these first ones. But because that would require speaking of several questions which are controversial among scholars, things I do not want to get mixed up with, I think it would be better to refrain from that and speak only in general about what these matters are, so that I leave it to wiser heads to judge if it would be useful for the public to be informed about more particular details. I have always 41 held firmly to the resolution that I have taken not to assume any principle other than the one which I have just used to demonstrate the existence of God and the soul, and to accept nothing as true which did not seem to me more clear and more certain than the proofs of geometers had seemed to me previously. Nonetheless, I venture to say that, not only did I find a way of satisfying myself in a short time concerning all the difficult principles which people are accustomed to dealing with in philosophy, but I also noticed certain laws which God has established in nature in such a way and of which He has impressed such notions in our souls that, after we have reflected on them sufficiently, we cannot doubt that they are precisely observed in everything which exists or which acts in the world. Then, as I considered the consequence of these laws, it seemed to me that I had discovered several truths more useful and more important than everything I had previously learned or even hoped to learn.

But since I attempted to explain these principal truths in a treatise which certain considerations prevented me from publishing,[27] I do not know how better to make them known than by stating here in summary form what that treatise contains. Before writing that text, I had intended to include in it all that I thought I knew concerning the nature of material things. But just as painters cannot portray equally well in a flat picture all the various surfaces of a solid body—they choose one of the main surfaces, which they set by itself facing the light and, by placing the others in shadows, do not 42 allow anything to appear more than one can see by looking at them—in the same way, fearing that I could not put in my discourse everything I had in my thoughts, I tried only to reveal there fairly fully what I understood

27 Here, Descartes refers to an earlier treatise titled *The World*, which wasn't published in its entirety until after Descartes's death. See the Introduction to Descartes section earlier in this volume.

about light, and then at the appropriate time, to add something about the sun and the fixed stars, because almost all light comes from them; about the heavens, because they transmit light; about the planets, comets, and the earth, because they reflect light; and in particular about all the bodies on earth, because they are colored, or transparent, or luminous; and finally about humans, because they are the ones who look at these things. Even so, in order to shade in all these things a little and to be able to speak more freely of what I was judging without being obliged to follow or to refute received opinions among the scholars, I resolved to leave everyone here to their disputes and to speak only of what would happen in a new world if God now created somewhere in imaginary space enough material to compose it, and if He set in motion, in a varied and disorderly way, the various parts of this material, so that it created a chaos as confused as poets could make it, and then afterwards He did nothing other than lend His ordinary help to nature and allow it to act according to the laws which He established.[28] Thus, first of all, I described this material and tried to picture it in such a way that there is nothing in the world, it seems to me, clearer and more intelligible, except what has been said from time to time 43 about God and the soul. For I even explicitly assumed that in the world there were none of those forms or qualities which people argue about in the schools, nor, in general, anything the knowledge of which was not so natural to our souls that we could not even pretend to remain ignorant of it. In addition, I made known the laws of nature, and without basing my reasoning on any principle other than the infinite perfections of God, I tried to demonstrate all of these laws about which one could entertain any doubts, to show that they are such that, although God could have created several worlds, there would not be one where these laws failed to be observed. After that, I showed how the greatest part of material in chaos would have to, as a result of these laws, organize and arrange itself in a certain way which made it similar to our heavens, and how, in so doing, some of its parts must have made up an earth and some parts planets and comets, and some other parts a sun and fixed stars. And at this point, dwelling on the subject of light, I explained at some length the nature of light which must be found in the sun and the stars, how from there it crossed in an instant the immense distances of heavenly space, and how it is reflected from the planets and comets towards the earth. To this I added several things concerning the material, the arrangement, the movements, and all the various

28 Descartes is describing a thought experiment in which he imagines how the world might have developed historically from material distributed randomly in the universe. The idea is potentially dangerous because it goes against the biblical description given in Genesis. Hence, later on Descartes explicitly denies that he is claiming the process he is summarizing actually took place.

qualities of these heavens and these stars. Consequently, I thought I had said enough about these matters to make known the fact that one observes nothing in these features of this world which must not, or at least could not, appear entirely similar to those of the world which I described. From 44 there I went on to speak in particular about the earth: about how, although I had expressly assumed that God had placed no heaviness in the material of which it is composed, all its parts could not help tending precisely to its center; how, having water and air on its surface, the arrangement of the heavens and the stars, and particularly of the moon, had to create on earth an ebb and flow similar in all its features to the ones we see in our oceans, and, beyond that, a certain flow in the water as well as in the air, from east to west, like the one we also observe between the tropics; how the mountains, seas, fountains, and rivers can naturally form out of that; how earth's metals come into the mines; how the plants on earth grow in the fields; and, in general, how all the things we call mixed or composite could be produced on earth. And, among other things, because I know of nothing, other than the stars, which produces light except fire, I studied to understand really clearly everything associated with the nature of fire: how it arises; how it is nourished; how sometimes it has heat without light and sometimes light without heat; how it can introduce various colors in different bodies, as well as various other qualities; how it melts some things and makes others harder; how it can consume almost everything or convert it into ash and smoke; and finally how, out of these cinders, simply by the violence of its actions, it makes glass. For this transformation of cinders into glass seemed to be as wonderful as anything else which happens in 45 nature, and I took particular pleasure in describing it.

However, I did not want to conclude from all these things that this world was created in the way I was proposing. For it is much more probable that God made the world from the beginning just what it had to be. But it is certain—and this is an opinion commonly accepted among theologians—that the actions by which God now preserves the world are exactly the same as the method by which He created it, in such a way that even if He did not give it at the start any form other than a chaos, providing that He had first established the laws of nature and had given His assistance so that it would act as it usually does, we can believe, without denying the miracle of creation, that because of these facts alone all purely material things would have been able, over time, to become the way we now observe them, and their nature is much easier to conceive when one sees them born gradually in this way than if one thinks of them only as made all at once in a finished state.[29]

---

29 Here Descartes is again coming close to potentially dangerous speculations. To propose that God's actions in developing the world are subject to natural

From the description of the inanimate bodies and of plants, I moved onto the bodies of animals and especially the body of humans. But because I did not yet have sufficient knowledge to speak of that in the same way as of other things—that is to say, to speak of effects in terms of causes, by revealing the seeds and the methods by which nature had to produce them—I contented myself with assuming that God formed the human body 46 completely like one of our own, both in the external shape of its limbs and in the arrangement of its inner organs, without making them of any material other than the one which I had described and without, at the outset, placing in that body any reasonable soul or any other thing to serve the body as a vegetative or sensitive soul,[30] except that He kindled in its heart one of those fires without light which I had already explained and which I conceived as being in no way different in its nature from the fire which heats hay when it is stored before it is dry or which makes new wines bubble when they are allowed to ferment on crushed grapes. For, by examining the functions which, as a result of this assumption, could be present in this body, I found precisely all those which could be in us without our being able to think, and thus those functions to which our soul—that is to say, that distinct part of the body whose nature is solely to think (as I have said above)—does not contribute, which are exactly the same as those in which we can say the animals without reason are similar to us. But in doing this, I could not find any of those functions which, because they are dependent on thought, are the only ones which pertain to us, to the extent that we are human; whereas I found all of them afterwards, once I assumed that God had created a reasonable soul and joined it to this body in the particular way which I described.

But so that you can see how I dealt with this material in that treatise, I want to include here the explanation for the movement of the heart and the arteries, the first and the most universal thing which one observes in

---

laws, even if God is the origin of those laws, is to suggest that there are some restrictions on God's later actions (i.e., His interventions and actions in the world must conform to those laws). However, the value of thinking about the development of the world as a historical process guided by laws (rather than as the product of the divine miracle of Creation) is that it enables human beings to come to a rational understanding of nature and thus makes the modern scientific study of nature possible, even if only in a thought experiment.

30 A reasonable soul is one that can think (reason), a vegetative soul can grow and reproduce, and a sensitive soul can have sensations. This categorization of types of soul goes back to Aristotle, and is intended to correspond to and explain the differences between humans, plants, and non-human animals. (Descartes later in this passage rejects Aristotle by denying here that there is any type of soul except the rational.)

animals. From that one will easily assess what one should think of all the others. And so that people have less difficulty understanding what I am 47 going to say, I would like those who are not versed in anatomy to take the trouble, before reading this, to have the heart of some large animal with lungs dissected in front of them. For it is in all respects sufficiently similar to the heart in humans. And I would like them to have demonstrated to them the two chambers or cavities which are in that heart. First, there is one chamber on its right side, to which two very large tubes correspond: that is, the *vena cava*, which is the principal receptacle of blood and, as it were, the trunk of the tree of which all the other veins of the body are the branches; and the *vena arteriosa*,[31] which has, with that label, been poorly named, because it is, in fact, an artery, the one which, originating at the heart, divides up, after leaving the heart, into several branches, which go out to distribute themselves throughout the lungs. Then there is the chamber on the left side of the heart, to which, in the same way, two tubes correspond, which are as large or larger than the ones just mentioned: that is, the venous artery, which is also misnamed, because it is nothing but a vein which comes from the lungs, where it is divided into several branches interwoven with those from the arterial vein and with those associated with the tube called the windpipe, through which air enters for respiration; and the large artery which, leaving the heart, sends its branches throughout the body. I would also like someone to point out carefully to them the eleven small strips of skin which, just like so many small doors, open and close the four openings in these two chambers, that is, three at the entry of the vena cava, where they are so arranged that they cannot in any way prevent the blood contained in the vena cava from flowing into the right chamber of the heart and, at the same time, effectively prevent its ability to flow out; three gates at the entry of the arterial vein, which, being arranged in 48 precisely the opposite way, easily allow the blood in this chamber to move toward the lungs but do not allow the blood in the lungs to return to that chamber of the heart. Then, in the same way, there are two other strips of membrane at the opening to the venous artery which allow the blood from the lungs to flow towards the left chamber of the heart, but prevent its return, and there are three at the entry of the great artery which allow blood to leave the heart but prevent it from returning there. There is no need to seek for any reason for the number of these membranes, beyond the fact that since the opening of the venous artery is an oval, because of its location, it can be readily closed with two; whereas, since the others are round, they can be more easily closed with three. In addition, I would like people to notice that the large artery and the arterial vein have a composition much harder and firmer than the venous artery and the vena cava; that these last

31 *vena arteriosa*: arterial vein, now called the pulmonary artery.

two get bigger before entering the heart and there form a structure similar to two small sacks, called the auricles of the heart, which are composed of flesh like that of the heart; that there is always more heat in the heart than in any other place in the body; and finally that, if any drop of blood enters its cavities, this heat in the heart is capable of making the drop quickly 49 swell and expand, just as all liquors generally do when one lets them fall drop by drop into some really hot container.

After all that, I have no need to say anything else to explain the movement of the heart, other than the following: when its cavities are not full of blood, then necessarily blood flows from the vena cava into the right chamber and from the venous artery into the left, because these two blood vessels are always full and their openings, which are oriented towards the heart, cannot then be blocked. But as soon as two drops of blood have entered the heart in this way, one in each of its chambers, these drops, which could only be of a considerable size because the openings through which they enter are very large and the vessels they come from are really full of blood, become thinner and expand, on account of the heat they encounter there, as a result of which they make the entire heart expand, and then they push against and close the five small gates which stand at the openings of the two vessels from which these drops of blood have come, thus preventing any more blood from moving down into the heart. And, continuing to become increasingly thinner, the drops of blood push against and open the six other small gates which stand at the opening of the two other vessels, through which they flow out, in this way causing all the branches of the arterial vein and great artery to expand, almost at the same instant as the heart, which immediately afterwards contracts, as do these arteries as well, because the blood which has entered them gets colder again there, and their six small gates close once more. Then the five valves on the vena 50 cava and the venous artery re-open, and allow passage of two more drops of blood, which, once more, make the heart and the arteries expand, just as in the preceding steps. And because the blood which enters the heart in this manner passes through these two small sacks called auricles, this motion causes the movement of the auricles to be the opposite of the heart's movement—they contract when the heart expands. As for the rest, so that those who do not understand the force of mathematical proofs and who are not accustomed to distinguishing true reasons from probable reasons do not venture to deny this matter without examining it, I wish to advise them that this movement which I have just explained is as necessarily a result of the mere arrangement of the organs which one can see in the heart with one's own eyes and of the heat which one can feel there with one's fingers and of the nature of blood which one can recognize from experience, as the movement of a clock is necessarily a result of the force, the placement, and the shape of its counter-weights and wheels.

But if someone asks how the blood in the veins does not exhaust itself as it flows continually into the heart in this way, and how the arteries are not overfilled because all the blood which passes through the heart goes into them, there's no need for me to say anything in reply other than what has already been written by an English doctor,[32] to whom we must give the honor of having broken the ice in this area and of being the first to teach that there are several small passages at the extremities of the arteries through which the blood which they receive from the heart enters into the small branches of the veins, from where it proceeds to move once again towards the heart, so that its passage is nothing other than a constant circulation. He proves this really well by the common experience of surgeons 51 who, having bound up an arm moderately tightly above a place where they have opened a vein, cause the blood to flow out more abundantly than if they had not tied the arm. And the opposite happens if they place the binding below the cut, between the hand and the opening, or if they make the binding above the opening very tight. For it is clear that the binding, when moderately tight, can only prevent the blood which is already in the arm from returning towards the heart by the veins, but in so doing the binding does not stop the blood from continuing to flow to the place from the arteries, because the arteries are situated below the veins and because the skin of the arteries, being harder, is less easy to press down. Thus, the blood which comes from the heart tends to move with more force through the arteries towards the hand than it does in returning from the hand towards the heart through the veins. And because this blood leaves the arm by the opening in one of the veins, it must necessarily be the case that there are some passages below this binding, that is to say, towards the extremities of the arm, through which it can come there from the arteries. He also demonstrates really well what he says about the flow of blood through certain small membranes which are so arranged in various places along the veins that they do not allow blood to move in the veins from the middle of the body towards the extremities, but only to return from the extremities towards the heart. Moreover, he demonstrates this by an experiment which shows that all the blood which is in the body can leave it in a very short time by a single artery, if it is cut, even if it has been tightly bound really close to the heart and cut between the heart and the binding, so that one simply could not imagine any explanation other than that the blood flowing 52 out is coming from the heart.

32 The English doctor is William Harvey (1578–1657), who published his pioneering work on the heart and circulation of the blood in 1628. Descartes accepted Harvey's view that the blood circulates, but rejected his idea that the heart is a pump, arguing that it caused circulation by heating and expanding blood.

But there are many other things which attest to the fact that the true cause of this movement of blood is as I have described it. For, firstly, the difference we notice between the blood which comes from the veins and the blood which flows out of the arteries could come about only if the blood is rarefied and, as it were, distilled in passing through the heart. It is more subtle, more lively, and warmer immediately after leaving the heart—that is to say, in the arteries—than it is shortly before entering the heart—that is to say, when it is in the veins. And if one pays attention, one will find that this difference is only readily apparent close to the heart and not so evident in places which are more distant from it. Secondly, the hardness of the skins making up the arterial vein and the large artery shows sufficiently well that the blood beats against them with greater force than it does against the veins. And why would the left chamber of the heart and the great artery be more ample and larger than the right chamber and the arterial vein, if it were not for the fact that the blood of the venous artery, which has only been in the lungs since passing through the heart, is more subtle and more strongly and more easily rarefied than the blood which comes immediately from the vena cava? And what could doctors diagnose by testing the pulse, if they did not know, in keeping with the fact that blood changes its nature, that it can be rarefied by the heat of the heart more or less strongly and more or less quickly than before? And if one examines how this heat is transferred to the other limbs, is it not neces-
53 sary to admit that it is by means of the blood, which, passing through the heart, is re-heated in it and from there spreads throughout the entire body? That's the reason why, if one takes blood from some part of the body, in that very process one takes the heat, and even if the heart were as hot as a burning fire, it would not be sufficient to re-heat the feet and the hands as much as it does, if it did not continually send new blood there. From this we also understand that the true purpose of respiration is to bring sufficient fresh air into the lungs to ensure that the blood which comes from the right chamber of the heart—where it has been rarefied and, as it were, changed into vapor—thickens and changes back again into blood, before falling back into the left chamber, without which it would not be fit to serve as nourishment for the fire there. What confirms this is that we observe that the animals which have no lungs also have only one cavity in the heart, and that children who cannot use their lungs while they are enclosed in their mother's womb have an opening through which blood flows from the vena cava into the left cavity of the heart and a passage by which the blood comes from the arterial vein into the large artery without passing through the lungs. Next, how would digestion take place in the stomach, if the heart did not send heat there through the arteries and with that some of the more easily flowing parts of the blood which help to dissolve the food which has been sent there? And the action which converts the juice of this

food into blood—surely that is easy to understand, if one considers that it is distilled, as it passes and re-passes through the heart, perhaps more than one or two hundred times each day? What else do we need to explain 54 nutrition and the production of the various humors[33] in the body, other than to say that the force with which the blood, as it gets rarefied, passes from the heart towards the extremities of the arteries, brings it about that some portions of it stop among those parts of the limbs where they are, and there take the place of some other parts which the blood pushes away, and that, depending on the situation or the shape or the smallness of the pores which these parts of blood encounter, some of them go off to certain places rather than to others, in the same way that anyone can see with various screens, which, being pierced in different ways, serve to separate various grains from one another? Finally, what is most remarkable in all this is the generation of animal spirits[34] which resemble a very slight wind or rather a very pure and very lively flame which, by climbing continually in great quantities from the heart into the brain, goes from there through the nerves into the muscles and gives movement to all the limbs, without it being necessary to imagine any other cause which has the effect of making the most agitated and most penetrating parts of blood, those most appropriate for making up these animal spirits, move towards the brain rather than elsewhere, other than that the arteries which carry these parts of the blood are those which come from the heart toward the brain by the most direct route and that, following the laws of mechanics—which are the same as nature's laws—when several things collectively tend to move towards the same place where there is insufficient room for all of them, as the parts of blood which leave the left cavity of the heart tend towards the brain, the most feeble and less agitated parts must be turned away from the brain by 55 the strongest parts. In this way, only the latter parts reach the brain.

I explained in particular detail all these things in the treatise which I had planned to publish previously. And then I demonstrated what the nerves and muscles in the human body must be made of, so that the animal spirits, once inside the nerves, would have the power to move its limbs, as one sees that heads, for a little while after being cut off, continue to move and bite the earth, in spite of the fact that they are no longer animated. I also showed what changes must take places in the brain to cause the waking state, sleep, and dreams; how light, sounds, smells, tastes, heat, and all the other qualities of external objects could imprint various ideas on the brain through the mediation of the senses; how hunger, thirst, and the other inner passions can also send their ideas to the brain; what must be understood

---

33 *humors*: fluids.

34 *animal spirits*: the very fine fluids that are thought to be carried by the nerves, and to explain their operation.

by common sense where these ideas are taken in,[35] by memory which preserves ideas, and by fantasy which can change them in various ways and compose new ones, and, in the same way, distribute animal spirits to the muscles and make the limbs of the body move in all the different ways—in relation to the objects which present themselves to the senses and in relation to the interior physical passions—just as our bodies can move themselves without being led by the will. None of this will seem strange to those who know how many varieties of *automata*, or moving machines, 56 human industry can make by using only very few pieces in comparison with the huge number of bones, muscles, nerves, arteries, veins, and all the other parts in the body of each animal. They will look on this body as a machine, which, having been made by the hand of God, is incomparably better ordered and more inherently admirable in its movements than any of those which human beings could have invented.

And here, in particular, I stopped to reveal that if there were machines which had the organs and the external shape of a monkey or of some other animal without reason, we would have no way of recognizing that they were not exactly the same nature as the animals; whereas, if there was a machine shaped like our bodies which imitated our actions as much as is morally possible, we would always have two very certain ways of recognizing that they were not, for all their resemblance, true human beings. The first of these is that they would never be able to use words or other signs to make words as we do to declare our thoughts to others. For one can easily imagine a machine made in such a way that it expresses words, even that it expresses some words relevant to some physical actions which bring about some change in its organs (for example, if one touches it in some spot, the machine asks what it is that one wants to say to it; if in another spot, it cries that one has hurt it, and things like that), but one cannot im- 57 agine a machine that arranges words in various ways to reply to the sense of everything said in its presence, as the most stupid human beings are capable of doing. The second test is that, although these machines might do several things as well or perhaps better than we do, they are inevitably lacking in some others, through which we would discover that they act, not by knowledge, but only by the arrangement of their organs. For, whereas reason is a universal instrument which can serve in all sorts of encounters, these organs need some particular arrangement for each particular action. As a result, it is morally impossible that there is in a machine's organs sufficient variety to act in all the events of life in the same way that our reason empowers us to act.

---

35 Common sense, in the Aristotelian tradition that Descartes is working within and modifying, is the faculty that receives our various sensations and unifies them into a single perceptual field.

Now, by these two same means, one can also recognize the difference between human beings and beasts. For it is really remarkable that there are no humans so dull and stupid, including even idiots,[36] that they are not capable of putting together different words and of creating out of them a conversation through which they make their thoughts known; by contrast, there is no other animal, no matter how perfect and how successful it might be, which can do anything like that. And this inability does not come about from a lack of organs. For we see that magpies and parrots can emit words, as we can, but nonetheless cannot speak the way we can, that is to say, giving evidence that they are thinking about what they are uttering; whereas people who are born deaf and dumb are deprived of organs which other people use to speak—just as much as or more than the animals—but 58 they have a habit of inventing on their own some signs by which they can make themselves understood to those who, being usually with them, have the spare time to learn their language. And this point attests not merely to the fact that animals have less reason than humans, but to the fact that they have none at all. For we see that it takes very little for someone to learn how to speak, and since we observe inequality among the animals of the same species just as much as among human beings, and see that some are easier to train than others, it would be incredible that a monkey or a parrot which was the most perfect of its species was not equivalent in speaking to the most stupid child or at least a child with a troubled brain, unless their soul had a nature totally different from our own. And one should not confuse words with natural movements which attest to the passions and can be imitated by machines as well as by animals, nor should one think, like some ancients, that animals speak although we do not understand their language. For if that were true, because they have several organs related to our own, they could just as easily make themselves understood to us as to the animals like them. Another truly remarkable thing is that, although there are several animals which display more industry in some of their actions than we do, we nonetheless see that they do not display that at all in many other actions. Thus, the fact that they do better than we do does not prove 59 that they have a mind, for, if that were the case, they would have more of it than any of us and would do better in all other things; it rather shows that they have no reason at all, and that it is nature which has activated them according to the arrangement of their organs—just as one sees that a clock, which is composed only of wheels and springs, can keep track of the hours and measure time more accurately than we can, for all our care.

After that, I described the reasonable soul and revealed that it cannot be inferred in any way from the power of matter like the other things I have

---

36 Descartes means by "idiot" here someone who is profoundly intellectually disabled.

spoken about, but that it must be expressly created, and I described how it is not sufficient that it be lodged in the human body like a pilot in their ship, except perhaps to move its limbs, but that it is necessary that the soul be joined and united more closely with the body, so that it has, in addition, feelings and appetites similar to ours and thus comprises a true human being.[37] As for the rest, here I went on at some length on the subject of the soul, because it is among the most important. For, apart from the error of those who deny God, which I believe I have adequately refuted above, there is nothing which distances feeble minds from the right road of virtue more readily than to imagine that the soul of animals is of the same nature as our own and that thus we have nothing either to fear or to hope for after this life any more than flies and ants do; whereas, once one knows how different they are, one understands much better the reasons which prove that the nature of our soul is totally independent of the body, and thus that it is not at all subject to dying along with the body. Then, to the extent that 60 one cannot see other causes which destroy the soul, one is naturally led to judge from this that the soul is immortal.

---

37 Descartes here announces one of the most challenging issue arising from his views. If the body is mechanical and the soul is not and if they must interact in some way, then how does that interaction take place? How can one explain consciousness in mechanistic terms? This remains a subject for debate in modern biology and the philosophy of mind.

PART SIX

It is now three years since I reached the end of the treatise which contains all these things and since I began to revise it in order to put it into the hands of a printer. I then learned that people to whom I defer, and whose authority over my actions could hardly be less than my own reason over my thoughts, had expressed disapproval of an opinion about physics published a little earlier by someone else.[38] I do not wish to say that I subscribed to that opinion, but, although I had observed in it nothing before their censure which I could imagine prejudicial to religion or the state, and thus nothing which would have prevented me from writing it if reason had persuaded me, this made me fear that there might nonetheless be something among my opinions where I had gone astray, notwithstanding the great care I always took not to accept new ideas into my beliefs for which I did not have very certain proofs and not to write anything which would work to anyone's disadvantage. This was sufficient to oblige me to alter my resolution to publish my opinions. For although the reasons I had adopted earlier had been very strong, my inclination, which has always led me to hate the profession of producing books, made me immediately find enough other reasons to excuse myself in this matter. And, given the nature of these reasons, on one side or the other, not only am I quite inter- 61 ested in stating them here, but the public may perhaps also be interested in knowing them.

I have never made a great deal of the things which come from my own mind, so while I gathered no other fruits from the method I was using other than that I satisfied myself concerning some difficulties in the speculative sciences or else that I tried to regulate my morals by reasons which my method taught me, I did not think myself obliged to write anything. For where morals are concerned, every person is so full of their own good sense that it would be possible to find as many reformers as heads, if it was permitted—to people other than those God has established as sovereigns over His people or those to whom He has given sufficient grace and zeal to be prophets—to try changing anything. Although my speculations pleased me a great deal, I thought that other people also had their own specula-

38 Descartes is here referring to Galileo, whose 1632 publication in defense of Copernicus' sun-centered model of the solar system got him into serious difficulties with the Church.

tions which pleased them perhaps even more. But immediately after I had acquired some general notions concerning physics and, by starting to test them on various particular difficulties, had noticed just where they could lead and how much they differed from principles which people have used up to the present time, I thought that I could not keep them hidden without sinning greatly against the law which obliges us to promote as much as we can the general good of all people. For my notions had made me see that it is possible to reach knowledge which is extremely useful for life, and that instead of the speculative philosophy which is taught in the schools, 62 we can find a practical philosophy by which, through understanding the force and actions of fire, water, air, stars, heavens, and all the other bodies which surround us as distinctly as we understand the various crafts of our artisans, we could use them in the same way for all applications for which they are appropriate and thus make ourselves, as it were, the masters and possessors of nature.[39] But it was not only a desire to invent an infinite number of devices which might enable us to enjoy without effort the fruits of the earth and all its amenities, but mainly also my desire for the preservation of our health, which is, without doubt, the principal benefit and the foundation of all the other benefits in this life. For even the mind depends so much on the temperament and the condition of the organs of the body that, if it is possible to find some means of making human beings generally wiser and more skilful than they have been up to this point, I believe we must seek that in medicine. It is true that the medicine now practiced contains few things which are remarkably useful. But without having any design to denigrate it, I am confident that there is no one, not even those who make a living from medicine, who would not claim that everything we know in medicine is almost nothing compared to what remains to be known about it, and that we could liberate ourselves from an infinity of illnesses, both of the body and the mind, and also perhaps even of the infirmities of old age, if we had sufficient knowledge of their causes and of all the remedies which nature has provided for us. Now, intending to 63 spend all my life seeking such a necessary science, and having encountered a road which seemed to me such that one should infallibly find this science by following it unless one was prevented either by the brevity of one's life or by the lack of experiments,[40] I judged that there was no better remedy against these two obstacles than to communicate faithfully to the public the little I had found and to invite good minds to try to go further by

39 In this famous statement Descartes makes clear one of the major purposes of the new natural philosophy (science): to gain power over nature.

40 The French word Descartes uses here, *expériences*, is sometimes translated merely as "observations," but Descartes means by it the kind of careful scientific observation that can be thought of as making an empirical experiment.

contributing—each according to their own inclination and power—to the experiments which need to be conducted and also by communicating to the public everything they learn, so that the most recent people begin where the previous ones have finished. If we thus joined the lives and labors of many people, collectively we might go much further than each particular person could.

Besides, I noticed that, where experiments are concerned, they are increasingly necessary as one's knowledge advances. At the beginning it is better to conduct only those which present themselves to our senses and which we cannot ignore provided that we engage in a little reflection, rather than to seek out more rare and recondite experiments, because these rarer ones are often misleading when we do not yet know the causes of the more common phenomena, and the circumstances on which they depend are almost always so particular and so precise that it is very difficult to observe them. But in this work I kept to the following order: first, I tried to find the general principles or the first causes of everything which exists 64 or could exist in the world, without considering anything germane to my purpose other than the fact that God alone created the world, not deducing anything additional other than certain grains of truth which are naturally in our souls. After that, I examined what were the first and most common effects we could deduce from these causes. In so doing, it seems to me, I found the heavens, the stars, and earth, and even, on the earth, water, air, fire, minerals, and some other things of the sort which are the most common of all and the simplest, and thus the easiest to know. Then, when I wanted to move down to more particular matters, so many varieties presented themselves to me that I did not think it would be possible for the human mind to distinguish the forms or species of bodies on the earth from an infinity of others which could exist there if the will of God had put them there and, thus, that one could not adapt them to our use unless one proceeded to the causes through the effects and made use of many particular experiments. After that, turning my mind to all the objects which had ever presented themselves to my senses, I venture to say that I didn't notice in them anything which I could not explain easily enough by the principles which I had found. But I must also confess that the power of nature is so ample and vast and its principles so simple and so general that I observed hardly any particular effect which I did not immediately understand as being capable of being deduced in several different ways, so that my greatest 65 difficulty is usually to find on which of these ways the effect depends. In dealing with this matter, I did not know any expedient other than, once again, to look for some experiments which would be such that their outcomes would not be the same if one of these ways had to explain it rather than some other way. As for the rest, I am now at a point where I perceive well enough, it seems to me, the method one must use to make the most of those experiments which can serve this purpose. But I also see that they are of such a kind and that there are so many of them that neither my hands nor my income, even if I had a thousand times more of both than I do, would suffice for all of them, so that from now on my progress in understanding nature will be proportional to the means I have of conducting more or fewer experiments. This was what I promised myself I would make known in the treatise which I had written, as well as showing in it the practical value which the public could gain from these experiments so clearly that I would oblige all those who wished to promote the general well being of humanity, that is to say, all those who are truly virtuous and who are not false by pretending to virtue or merely virtuous by public opinion, to communicate to me all the experiments which they have already made, as well as to help me in researching those which remain to be done.

But since that time I have had other reasons which made me change my mind and think that I really must continue to write down all matters which I judged to have some importance, to the extent that I discovered truth in them, and to bring to my writings the same care that I would if I wanted 66 them published, so that I would have more time to examine such things well, since there is no doubt one always looks more closely at what one thinks many people must see than at what one does only for oneself, and often matters which seemed to me true when I began to think of them appeared to me false when I wished to put them on paper. By writing things down, I would not lose any opportunity to benefit the public, if I could, and, if my writings are worth anything, those who have them after my death could use them wherever they were most relevant. But I thought I must not, on any account, agree that they be published during my lifetime, so as to prevent the hostility and controversies which they could perhaps arouse—and the reputation which I could acquire—from giving me any occasion to waste the time which I planned to use to instruct myself. For although it is true that each person is obliged to provide as much as is in them for the good of others, and there is no value whatsoever in anything which has no benefit for anyone, nevertheless it is also true that we should care about things more distant than the present and that it is good to omit things which might bring some benefit to those now living when one's intention is to create other things which will bring more benefits to our descendants. In fact, I really wanted people to understand that the little I had learned up to this point is almost nothing in comparison with what I am ignorant of and what I do not despair of being able to learn. For it is almost the same with those who discover truth little by little in the sciences 67 as it is with those who, once they start to become rich, have less trouble in making large acquisitions than they did previously, when they were poorer, in making much smaller ones. Or, again, one can compare them to leaders of armies whose forces usually grow in proportion to their victories and who, in order to capture towns and provinces, need more leadership to maintain their forces after losing a battle than they do after winning one. For it is truly a matter of waging battles when one tries to overcome all the difficulties and mistakes which prevent us from reaching an understanding of the truth. And it is a defeat in battle when one accepts some false opinion concerning any general and important matter. Afterwards one requires a great deal more skill to put oneself in the same condition one was in previously than one requires to make great progress when one already has confirmed principles. In my case, if I have previously found some truths in the sciences (and I hope that the matters contained in this volume will make people conclude that I have found some), I can say that those are only the consequences of and dependent upon five or six major difficulties which I overcame, and that I count these as so many battles in which victory was on my side. Still, I will not hesitate to state that I think I need to win only two or three others like those in order to reach the final goal of my project and that I am not so advanced in years that, given the ordinary course of nature, I could not still have enough leisure to bring my project to its conclusion. But I think that I am all the more obliged to better [68] manage the time remaining to me, now that I have more hope of being able to use it well, and I would no doubt have many chances to lose that time if I published the foundations of my physics. For although these foundations are almost all so evident that one need only hear them to believe them and there are none of them which, in my view, I cannot demonstrably prove, nevertheless, because it is impossible that they will agree with all the various opinions of other people, I anticipate being often distracted by the hostility they would give rise to.

One could say that this opposition would be useful, to the extent it makes me understand my mistakes, and that, if I have anything good to say, others will by this means have a more complete understanding. Since several people can see more than one person by themselves, if people begin from now on making use of my principles, they will also help me with their inventions. But even though I recognize that I am extremely subject to error and that I almost never have faith in the first thoughts which come to me, nevertheless the experience which I have of objections which people could make about me prevents me from hoping they will yield any benefit. For I have already borne the criticism of so many of those whom I held as friends and of some others who I thought considered me indifferently and even of some in whom I knew malignity and envy would try hard enough to uncover what affection concealed from my friends. But rarely has someone made an objection which I had not in some way anticipated, unless it was really distant from my subject, so that I have almost never [69] met any critic of my opinions who did not appear to me to be less rigorous or less fair than myself. Moreover, I have never observed that anyone has discovered any truth of which people were previously ignorant by means of the disputes practiced in the schools. For when each person tries to emerge victorious, people strive much harder to establish probability than to weigh the reasons on one side or the other, and those who have long been good lawyers are not, on that account, better judges afterwards.

As for the practical use which other people derive from the communication of my thoughts, it could not be all that great, since I have not taken them so far that there is no need to add a great many things before they can be practically applied. And I think I can say without vanity that if there is anyone who can do that, this person should be me rather than anyone else, not because many minds incomparably better than mine could not be found in this world, but because one cannot conceive of something so well and make it one's own when one learns it from someone else as when one comes up with it oneself. What is really true about this matter is that, although I have often explained some of my opinions to people with very good minds, who, while I was speaking to them, seemed to understand my opinions very clearly, nonetheless, when they have repeated them, I have noticed that almost always they have changed them in such a way that I could no longer admit them as mine. Incidentally, I am more than happy to 70 take the opportunity here to beg our descendants never to believe anything that people tell them comes from me which I have not divulged myself. I am not astonished at the extravagant things which people attribute to those ancient philosophers whose writings we do not possess. Nor do I judge that their thoughts were really irrational on that account, seeing that they were the best minds of their times, but merely assume that their thoughts have been misrepresented to us. For we see also that it almost never happens that any of their disciples surpasses them, and I am confident that the most passionate of those people who follow Aristotle nowadays would consider themselves fortunate if they could have as much knowledge of nature as he had, even on condition that they would never know any more. They are like ivy which tends not to climb higher than the trees which support it and which often even comes down again when it has reached the tree tops. For it seems to me that those people also come back down, that is, make themselves in some way less knowledgeable than if they were to abstain from studying, when, not content with knowing everything intelligibly explained by their author, they wish to find, beyond that, the solution to several difficulties about which they have said nothing and have perhaps never even thought. However, their way of practicing philosophy is extremely comfortable for those who have nothing but really mediocre minds, for the obscurity of the distinctions and the principles they use enables them to speak of everything as boldly as if they understood what they were talking 71 about and to defend everything that they state against the most subtle and skillful minds, without anyone having the means to argue against them. In this it strikes me they are similar to a blind person who, in order to fight on equal terms against someone who can see, makes them come deep into some really dark cave. And I can state that such people have an interest in my abstaining from publishing the principles of philosophy I use, because, given that they are very simple and very evident, if I published them I would be doing roughly the equivalent of opening some windows and bringing the light of day into this cave where they have gone down to fight each other. But even the best minds have no occasion to want to know these principles. For if they want to know how to speak about everything and to acquire the reputation of being scholarly, they will get there more easily in contenting themselves with probability which can be found in all sorts of matters without great trouble, rather than by seeking out the truth, which is not discovered except little by little in some matters and which, when it is a question of speaking of other matters, requires us to confess frankly that we are ignorant of them. If they would rather have the undoubtedly preferable condition of knowing a few truths over the vanity of appearing to be ignorant about nothing, and if they wish to follow a plan similar to my own, they do not need me to say anything more to them than I have already said in this discourse. For if they are capable of moving on further than I have done, they will also, with all the more reason, find for themselves everything I think I have found. Since I have never looked at anything except in due order, it is certain that what remains for me to discover is inherently more difficult and more hidden than what I have been 72 able to find up to this point, and they would have much less pleasure in learning that from me than from themselves. Beyond that, the habit they will acquire by searching first for easy things and then moving on gradually by degrees to other more difficult things will serve them better than all my instructions would be able to. As for me, I am convinced that had I been taught from youth all the truths for which I have since sought demonstrations, and had I had no trouble learning them, I would perhaps have never known any others. At the very least I would never have acquired the habit and the skill which I think I have in constantly finding new truths to the extent that I apply myself in looking for them. In a word, if there is in the world some work which cannot be properly completed by anyone other than the same person who started it, it's the work I do.

It is true that, in terms of the experiments which can help in this work, one person by themselves is not sufficient to undertake them all. But they cannot put to practical use hands other than their own, except those of artisans or such people as they can pay, who in hopes of profit—which is a very effective means—would carry out everything exactly as instructed. As for volunteers who from curiosity or a desire to learn perhaps offer to help—apart from the fact that ordinarily they promise more than they deliver and that they come up with nothing but fine proposals, none of

73 which ever succeeds—they inevitably want to be paid with the explanation of some difficulties or at least with compliments and useless discussions which would cost a person more time than they could afford. As for the experiments which other people have already carried out, even though they should be willing enough to speak about them, those who call such experiments secrets would never do so. Such experiments for the most part contain so many superfluous circumstances or ingredients that it would be very difficult for one to decipher their truth. Beyond that, one would find almost all of them so badly explained or even false—because those who have carried them out have forced themselves to make the experiment appear to conform to their principles—that if there were some experiments one could use, once again it would not be worth the time needed to pick them out. In the same way, if there were in the world someone whom people were sure would be capable of finding the greatest and most useful things for the public as possible, and if for this reason other people tried hard to help them in every way to carry out their project with success, I don't see that they could do anything for them other than furnish the costs of the experiments they needed to carry out and, as to the rest, prevent their leisure from being taken away by anyone's importunity. But beyond the fact that I do not presume so much of myself as to wish to promise anything extraordinary and that I do not indulge in thoughts so vain as to imagine to myself that the public ought to show a great deal of interest in my plans, I do not have a soul so base that I would be willing to accept 74 from anyone a favor which people might think I did not deserve.

All these considerations combined were the reason, three years ago, that I did not wish to publish the treatise which I had in my hands. I even made a resolution that during my lifetime I would not make public any other treatise which was so general and from which one could learn the foundations of my physics. However, once again, there have been two other reasons since then which have obliged me to set down here some particular essays and to give the public some account of my actions and my plans. The first is that if I failed to do this, several people who knew of my previous intention to have some of my writings published could imagine that the reasons why I held back from doing so were more disadvantageous to me than they were. For although I do not like glory excessively, or, dare I say it, although I dislike it to the extent that I see it as contrary to peace and quiet, which I value above everything, nonetheless I have also never tried to hide my actions as if they were crimes, nor have I taken many precautions to remain unknown, as much because I would have thought it would harm me if I did so, as because it would have given me a sort of unease which would, once again, have been contrary to the perfect peace of mind which I am looking for. Also, being always indifferently poised between caring to get known or not to get known, since I could not prevent myself from acquiring some kind of reputation, I thought that I ought to do my best at least to avoid having a bad one. The other reason which obliged me to write this is that I realized more and more every day how the plan I 75 have to teach myself was suffering a delay because of an infinite number of experiments I needed and because it is impossible to carry them out without the help of others. Although I do not flatter myself so much as to hope that the public pays great attention to my interests, nonetheless by the same token I do not want to let myself down so much as to provide an excuse for those who will come after me to reproach me some day by saying that I could have left many things much better than I did, had I not so neglected making them understand the ways in which they could contribute to my project.

And I thought that it would be easy for me to choose some matters which, without being subject to a great deal of controversy and without requiring me to state more of my principles than I wanted to, would permit me to reveal with sufficient clarity what I could or could not do in the sciences. In these matters I cannot say if I have been successful, and I have no desire to ward off anyone's criticism as I speak in person about my own writings. But I will be very pleased if people examine them. In order for people to have more chances to do this, I request that all those who could have some objections take the trouble to send them to my publisher. If he tells me about them, I will try to attach my response to their objection and publish them at the same time. In this way, the readers, seeing the objections and my replies together, will judge the truth all the more easily. For I promise never to make long replies to such objections, but only to confess my errors very candidly, if I recognize them, or else, if I cannot see them, 76 to state simply what I believe is required to defend the things I have written, without adding there an explanation of any new material, so as not to get endlessly involved with one matter after another.

If some of those things which I have spoken about at the beginning of *On Dioptrics* and *On Meteors*[41] are shocking at first, because I call them suppositions and I do not seem to have any desire to prove them, I urge people to have the patience to read the whole text attentively, and I hope that they will be satisfied with it. For it seems to me that the reasons follow there in sequence in such a way that the last ones are established by the first ones, which are their causes, and the first ones are reciprocally established

41 These are the essays, published with the *Discourse*, that are usually referred to as the *Optics* (which examines the properties of light, the eye, and lenses) and the *Meteorology* (a naturalistic account of various atmospheric conditions, such as weather—the word "meteorology" comes from ancient Greek words meaning "the study of things high in the air," which is also from where the English word "meteor" derives).

by the last ones, which are their effects. And people should not imagine that, in doing this, I am committing the error which logicians call arguing in a circle. For since experimentation makes most of these effects very certain, the causes which I have deduced do not serve so much to prove these effects as to explain them, so the case is precisely reversed: it is the causes which are proved by the effects. And I used the name suppositions for these causes so that people might know that I think I can deduce them from these first truths which I have explained above, but that I expressly wanted to avoid doing so in order to prevent certain minds who imagine that they understand in a single day everything that another person has thought out in twenty years as soon as that person has said only two or three words about these matters to them, and who are all the more subject to error and less capable of truth the more penetrating and bright they are, 77 from taking the opportunity to construct some extravagant philosophy on what they believe are my principles, and in order to prevent people blaming me for that. As for the opinions which are entirely mine, I do not seek to excuse them as new. To the extent that people think carefully about the reasons for them, I am confident that they will find my opinions so simple and so consistent with common sense that they will seem less extraordinary and less strange than some others which people might have on the same subjects. In addition, I do not boast that I am the first inventor of any of them, although I have never accepted them merely because they were said by others or because they have not been said by others, but simply because reason persuaded me to accept them.

If artisans cannot immediately carry out the inventions explained in the *Dioptrics*, I do not think that people can, for that reason, say that the text is a poor one. For to the extent that dexterity and skill are required to make and to adjust the machines which I have described without missing the slightest detail, I would be no less amazed if they were successful on the first attempt than if someone could learn in a single day to play the lute extremely well simply because someone had given them a good musical score. And if I write in French, the language of my country, rather than in Latin, the language of my teachers, the reason is that I hope those who use only their natural reason, pure and simple, will judge my opinions better than those who believe nothing but ancient books.[42] And as for those who combine good sense with study, who are the only ones I hope to have as

42 Descartes's choice to write in French rather than the scholarly language of Latin, unusual for a natural philosopher of the time, did increase his readership in France but also—Descartes thought—harmed his exposure to his scientific colleagues. As a consequence, his subsequent writings were in Latin and he agreed to have the *Discourse* published in a Latin translation a few years later (in 1644).

my judges, I am confident that they will not be so partial to Latin that they 78 will refuse to listen to my reasons because I explain them in the common language.

As for the rest, I do not wish to speak here in particular detail about the future progress which I hope to make in the sciences, nor to commit myself to promising the public what I am not confident of achieving. But I will only say that I have resolved not to use the time remaining to me for anything other than trying to acquire some knowledge of nature of such a kind that people can derive from it rules for medicine more reliable that those which they have at present, and that my inclination keeps me so far away from all kinds of other projects—mainly those which can be practically useful to some people only by harming others—that if some circumstance forced me to use my time in this way, I do not think I would be capable of succeeding in it. In saying this, I am making a declaration here which I well understand cannot make me important in the world, but also I have no desire to be important. I will always hold myself more obliged to those by whose favor I enjoy my leisure unencumbered than to those who might offer me the most prestigious positions on earth.

MEDITATIONS ON FIRST PHILOSOPHY

What Is the Structure of the *Meditations*?

The *Meditations* is not intended to be merely an exposition of philosophical arguments and conclusions, but is supposed to be an exercise in philosophical reflection for the reader. As Bernard Williams has put it, "the 'I' that appears throughout them from the first sentence on does not specifically represent [Descartes]: it represents anyone who will step into the position it marks, the position of the thinker who is prepared to reconsider and recast his or her beliefs, as Descartes supposed we might, from the ground up."[1] Descartes aims to convince us of the truth of his conclusions by making us conduct the arguments ourselves. (It is interesting to note that the structure of the *Meditations* was modeled on the "spiritual exercises" that students at Jesuit schools, such as the one Descartes attended, were required to undertake in order to learn to move away from the world of the senses and to focus on God.)

In the First Meditation the thinker applies a series of progressively more radical doubts to his or her preconceived opinions, which leaves her unsure whether she knows anything at all. But then in the Second Meditation the thinker finds a secure foundational belief in her indubitable awareness of her own existence. The rest of this meditation is a reflection on the thinker's own nature as a "thinking thing." In the Third Meditation the thinker realizes that final certainty can only be achieved through the existence of a non-deceiving God, and argues from the idea of God found in her own mind to the conclusion that God must really exist and be the cause of this idea (this is sometimes nicknamed the "Trademark Argument," from the notion that our possession of the idea of God is God's "trademark" on his creation). The Fourth Meditation urges that the way to avoid error in our judgments is to restrict our beliefs to things of which we are clearly and distinctly certain. The Fifth Meditation introduces Cartesian science by discussing the mathematical nature of our knowledge of matter, and also includes a second proof for God's existence which resembles the eleventh-century "ontological argument" of St. Anselm. Finally, in the Sixth Meditation, the thinker re-establishes our knowledge of the real existence of the

1 This appears in his introductory essay to the Cambridge University Press edition of the *Meditations* (1996).

external world, argues that mind and body are two distinct substances, and reflects on how mind and body are related.

Some Useful Background Information

1. Descartes makes frequent use of the terms "substance," "essence," and "accident." A substance is, roughly, a bearer of attributes, i.e., a thing that has properties, or what the properties are properties *of*. The essence of a substance is its fundamental intrinsic nature, a property without which that thing, of that sort, could not exist. Descartes held that for every substance there is exactly one property which is its essence. A substance's "accidents"[2] are all the rest of its properties, the ones which are not part of its essence.

   Take, for example, a red ball: its redness, the spherical shape, the rubbery feel, and so on, are all properties—accidents—of the ball, and the ball's substance is the "stuff" that underlies and possesses these properties. According to Descartes, the fundamental nature of this stuff—its essence—is that it is extended in three dimensions, that it fills space.

   For Descartes and his contemporaries there is also another important aspect to the idea of substance. Unlike an instance of a property, which cannot exist all by itself (there can't be an occurrence of redness without there being something which is red—some bit of substance which is the bearer of that property), substances are not dependent for their existence on something else. In fact, for Descartes, this is actually the *definition* of a substance: "By 'substance' we can understand nothing other than a thing which exists in such a way as to depend on no other things for its existence" (*Principles of Philosophy*). So, for instance, a tree is not really a substance, since trees do depend for their existence on other things (such as soil, light, past trees, and so on). On the other hand, according to Descartes, matter itself—all matter, taken as a whole—is a substance. Matter cannot be destroyed or created (except by God), it can only change its local form, gradually moving from the form of a tree to the form of a rotting tree trunk to the form of soil, for example.

2. Descartes relies quite heavily on a contrast between understanding, the will, and imagination. The basic distinction is as follows: the un-

2 Strictly speaking, Descartes thought of accidents as being 'modes' of the one essential property of the substance (rather than being really separate properties): shape, for example, is a mode of being extended in space.

derstanding (sometimes translated as "intellect") and the will are the two primary faculties of the mind. The understanding is our intellectual apprehension of ideas, the faculty by which the mind considers the contents of thoughts (indeed, without understanding, mere sensations have no content at all). The will is our ability to either assent to or reject these ideas—it is our faculty of judgment. An act of will (assent) is necessary for there to be a belief at all, according to Descartes—a mere idea by itself can be neither accurate nor erroneous. Imagination[3] is a faculty that—like sensation—is dependent on the existence of one's physical body. It involves the presentation of ideas, especially mental images. These images, though, are not mere two-dimensional 'mental pictures,' as we might assume; for Descartes they are certainly not always visual, and are also probably in some sense extended objects (and hence as much physical as mental).

Two key details are worth bearing in mind for a fuller appreciation of Descartes's arguments. First, although the understanding/imagination distinction would have been quite familiar at the time, Descartes departs from the intellectual tradition of Aristotle in holding that sensation and imagination, though they certainly intimately involve the body, are faculties of the mind. (Aristotle and his followers held that only the intellect is properly mental.) Furthermore, Descartes emphasized that our sensations, by themselves, tell us nothing about the world—only our understanding generates judgments.

Second, Descartes believed that the understanding is a passive, rather than an active, mental faculty: it takes in the deliveries of sensation, for example, and produces thoughts almost automatically, according to the way in which God has created us. By contrast, the will (also given to us by God) is an active mental faculty. We cannot choose how the world appears to us, but we can choose whether or not to certify those judgments as accurate. In this way, for Descartes, error is almost a moral failing—a failure of the will. Our mistakes are not God's fault, but our own.

3. A related phrase frequently used by Descartes in the *Meditations* is "natural light." Descartes has in mind here what in earlier writings he calls "the light of reason"—the pure inner light of the intellect,

---

3 For Descartes and his contemporaries, "imagination" is meant in a more particular sense than the modern one. Today "imagination" includes any kind of speculation or invention, while Descartes's use of the term continues the earlier usage, meaning specifically the having of mental images.

a faculty given to us by God, which allows us to see the truth of the world much more clearly than we can with the confused and fluctuating testimony of the senses.

4. Descartes, following the scholastic jargon of the time, calls the representational content of an idea its "objective reality" (he uses this term in his attempted proof of the existence of God in the Third Meditation). Confusingly, for something to have merely objective reality in this sense is for it to belong to the mental world of ideas, and not to the mind-independent external world at all.

   For example, if I imagine Santa Claus as being fat and jolly, then Descartes would say that fatness is "objectively" present in my idea—an idea of fatness forms part of my idea of Santa Claus. By contrast, the baby beluga at the Chicago aquarium is fat, but its fatness is not merely the *idea* of fatness but an actual property of the beluga. In general, for any idea *I* that represents a thing *X* which has the property of being F, F-ness will be present formally in *X* but objectively in *I*.

   What Descartes and his contemporaries call "formal reality," then, is just the reality something has simply by virtue of existing. Since ideas exist (they are modes of a thinking substance), they have both formal reality—the idea itself—and objective reality—the content of the idea.

5. Descartes's talk of a non-physical "soul" was in accord with contemporary Christian theology. However, his reasons for holding that the mind is immortal and non-material (given largely in the Second and Sixth Meditations) were not primarily religious ones. For Descartes the word "soul" simply means the same as the word "mind," and encompasses the whole range of conscious mental activity, including the sensations of sight, touch, sound, taste, and smell; emotions (such as joy or jealousy); and cognitive activities like believing, planning, desiring, or doubting. For Descartes the mind, or soul, is also to be distinguished from the brain: our brains, since they are extended material things, are part of our body and not our mind.

Some Common Misconceptions

1. The *Meditations* describe a process in which the thinker moves from pre-reflective starting points towards a clearer understanding of knowledge. As a consequence, not everything that Descartes writes—especially in the earlier meditations—is something that he, or the thinker, will agree with by the time they have completed the

process. (For example, he begins by saying that "[u]p to this point, what I have accepted as very true I have derived either from the senses or through the senses," which is a principle he later rejects.)

2. Descartes is not a skeptic. Although he is famous for the skeptical arguments put forward in the First Meditation, he uses these only in order to go beyond them.[4] It is a bit misleading, however, to think of Descartes as setting out in the *Meditations* to defeat skepticism: his main interest is probably not in proving the skeptic wrong, but in discovering the first principles upon which a proper science can be built. He uses skepticism, surprisingly, in order to create knowledge—to show that a properly constituted science would have nothing to fear from even the most radical doubts of the skeptic. Thus, for example, Descartes does not at any point argue that we should actually believe that the external world does not exist—instead, he suspends his belief in external objects until he has a chance to properly build a foundation for this belief (and by the end of the *Meditations* he is quite certain that the external world exists).

3. Although "I think; therefore, I am" is the first step in Descartes's reconstruction of his knowledge in the *Meditations*, it is nevertheless not the first piece of *knowledge* that he recognizes—it does not arise out of a complete knowledge vacuum. Before the thinker can come to know that "I think; therefore, I am" is true, Descartes elsewhere admits that they must know, for example, what is meant by thinking, and that doubting is a kind of thought, and that in order to think one must first exist. Therefore, it is best to think of "I think; therefore, I am," not as the first item of knowledge, but as the first *non-trivial* piece of secure knowledge about the world that a thinker can have. It's a piece of information not just about concepts or logic but actually about the world—but, according to Descartes, it's information we can only get if we *already* (somehow) possess a certain set of concepts.

4. Descartes's project is to show how, by setting our knowledge of the world on firm foundations, we can overcome any skeptical doubts and have confidence in the conception of reality that results. It is

---

4 In one of his letters, Descartes noted that, when ancient medical authorities such as Galen or Hippocrates wrote about the causes of disease, no one accused them of telling people how to get sick; in the same way, Descartes complained, "I put forward these reasons for doubting not to convince people of them but, on the contrary, in order to refute them."

less widely appreciated, however, that the common-sense picture of the world, as it is apparently revealed to our senses, with which Descartes begins is very different from the worldview with which we are left at the end of the *Meditations*. Descartes does not merely rescue common sense from skepticism; instead, he replaces a naïve view of the world with a more modern, scientific one.

5. Descartes is sometimes portrayed as making the following (bad) argument to establish that mind is distinct from body: I can doubt that my body exists; I cannot doubt that my mind exists; there is therefore at least one property that my mind has which my body lacks (i.e., being doubtable); and therefore mind and body are not identical. Descartes does seem to make an argument which resembles this (in the Second Meditation), but he later denied that this is really what he meant, and he formulates much stronger—though perhaps still flawed—arguments for dualism in the Sixth Meditation.

Suggestions for Critical Reflection

1. Descartes, in the *Meditations*, has traditionally been seen as raising and then trying to deal with the problem of radical skepticism: that is, according to this interpretation, he raises the possibility that (almost) all our beliefs might be radically mistaken and then argues that this is, in fact, impossible. A more recent line of interpretation, though, sees Descartes not as attempting to answer the skeptic, but as trying to replace naïve empirical assumptions about science with a more modern, mathematical view—in particular, that Descartes is trying to show our most fundamental pieces of knowledge about mind, God, and the world come not from sensory experience, but directly from the intellect. Which interpretation do you think is more plausible? Could they both be right? If Descartes does want to refute skepticism, is he successful in doing so? If his goal is to overturn naïve scholastic empiricism, do you think he manages to do that?

2. Descartes's foundational claim is "I think; therefore, I am." How does Descartes justify this claim? Does he have, or need, an *argument* for it? Is an argument that justifies this claim even possible?

3. Descartes writes, in the Third Meditation, "[t]here cannot be another faculty [in me] as trustworthy as natural light, one which could teach me that the ideas [derived from natural light] are not true." What do you think he means by this? Is he right? How important to his arguments is it that he be right about this?

4. Eighteenth-century Scottish philosopher David Hume dryly said, of the *Meditations*, "To have recourse to the veracity of the supreme Being, in order to prove the veracity of our senses, is surely making a very unexpected circuit" (Section XII of *An Enquiry Concerning Human Understanding*, 1748). What do you think? Does Descartes establish the existence of God?

5. On Descartes's picture, do you think an atheist can have any knowledge? Why or why not?

6. It seems to be crucial to Descartes's arguments (especially those in the Fourth Meditation) that God is not responsible for our errors, that what we believe—and, indeed, whether we believe or simply suspend our belief—is something that is under our direct control: that we can freely will to believe or not. Does this seem plausible to you? (Could you really decide not to believe that, say, your body exists?) How might Descartes argue for this position? If it cannot be defended, how problematic would this be for the project of the *Meditations*?

7. A famous objection to Descartes's conclusions in the *Meditations* (raised for the first time by some of his contemporaries) is today known as the problem of the Cartesian Circle. Descartes says in the Third Meditation, "Whatever I perceive very clearly and distinctly is true." Call this the CDP (Clear and Distinct Perception) Principle. It is this principle that he thinks will allow him to reconstruct a body of reliable scientific knowledge on firm foundations. However, he immediately admits, the CDP Principle will only work if we cannot ever make mistakes about what we clearly and distinctly perceive; to show this, Descartes tries to prove that God exists and has created human beings such that what we clearly and distinctly see to be evidently true really is true. But how does Descartes prove God exists? Apparently, by arguing that we have a clear and distinct idea of God, and so it must be true that God exists. That is, the objection runs, Descartes relies upon the CDP Principle to prove that the CDP Principle is reliable—and this argument just goes in a big circle and doesn't prove anything. What do you think of this objection?

8. "How do I know that I am not ten thinkers thinking in unison?" (G.E.M. Anscombe, "The First Person" [1975]). What, if anything, do you think Descartes has proved about the nature of the self?

9. How adequate are Descartes's arguments for mind–body dualism? If mind and body are two different substances, do you think this might cause other philosophical problems to arise? For example, how might mind and body interact if they are radically different and have no properties in common? How could we come to know things about other people's minds? How could we be sure whether animals have minds or not, and if they do what they might be like?

10. Descartes recognized no physical properties but size, shape, and motion. Where do you think Descartes would say colors, tastes, smells, and so on come from?

RENATI

DES-CARTES,

MEDITATIONES DE PRIMA

PHILOSOPHIA,

IN QVA DEI EXISTENTIA ET ANIMÆ IMMORTALITAS DEMONSTRATVR.

PARISIIS,
Apud MICHAELEM SOLY, viâ Iacobeâ, sub signo Phœnicis.

M. DC. XLI.

Cum Priuilegio, & Approbatione Doctorum.

Meditations on First Philosophy

in which the existence of God and the difference between the human soul and body are demonstrated[1]

To the very learned and most illustrious Dean and Doctors of the Sacred Faculty of Theology in Paris[2]

1

The reason urging me to offer you this treatise is compelling, and I believe that, once you understand the organizing principle of what I have undertaken, you will have an equally just reason for taking the work under your protection. Such is my confidence, in fact, that I can find no better way of recommending it to you than by outlining briefly what I set out to do in it.

Of those questions which ought to be resolved with the help of Philosophy rather than of Theology, I have always thought that the two most important concerned God and the soul. For although among those of us who believe, faith is sufficient to accept that the human soul does not perish with the body and that God exists, it really does not seem at all possible to 2

1 The subtitle for the first edition was "In which the existence of God and the immortality of the soul are demonstrated," but Descartes was dissatisfied with it, and in the second Latin edition, published a year after the first, the subtitle was changed to the one used here, which more accurately reflects the content of his argument. The original title page also includes (at the bottom) the phrase "With the Official Sanction and Approval of the Scholars" (*Cum Privilegio, et Approbatione Doctorum*), an odd addition since the *Meditations* opens with a plea to the Sorbonne to grant him such a favor. It is unclear whether the learned doctors of the Sorbonne ever gave the work their official approval. The title page of the first French edition follows the second Latin edition and adds "And the objections made against these Meditations by various very scholarly people, with the author's responses." The Latin and French editions spell the author's name Des Cartes or Des-Cartes.

2 The University of Paris, one of the first European universities, was founded in the twelfth century; the Collège de Sorbonne, one of its theological colleges, was established in 1257. By the time that Descartes was writing it had become the most distinguished theological institution in France.

convince non-believers about any religion, and perhaps about any moral virtue as well, unless one first establishes the truth of those two questions for them by natural reason. And since this life frequently offers greater rewards for vice than for virtue, few people would prefer what is right to what is convenient, if they did not fear God and were not anticipating a life hereafter. It is indubitably true that we must believe in the existence of God because that is what we are taught by the Holy Scriptures; and that, on the other hand, we must believe the Holy Scriptures because they come from God (for since faith is a gift from God, obviously the same Being who gives us grace to believe other things can also give us grace to believe in His own existence). However, we cannot make this argument to non-believers, for they would claim that such reasoning is circular. And, indeed, I have observed not only that all of you, as well as other theologians, affirm that the existence of God can be proved by natural reason, but also that it can be inferred from sacred Scripture, that we can acquire knowledge of Him [much] more readily than of many created things, and that, in fact, it is so utterly easy that those who lack such knowledge are themselves to blame, as we can see from these words in Wisdom 13: *And these men ought not to be forgiven, for if they could know so much that they were able to assess the things of this world, why did they not find the Lord of these things more easily?*[3] And Romans, Chapter 1, states that such men *have no excuse*.[4] And, once again, in the same place, the following words *What is known of God is manifest in them* appear to be advising us that everything which can be known about God can be revealed by reasons we do not derive from anywhere other than our own minds. Thus, I did not think it would be inappropriate for me to explore how that might be done and by what road God might be known more easily and more certainly than worldly matters.

3 As far as the soul is concerned, many people have judged that its nature cannot be investigated easily, and some have even dared to claim that human reasoning has convinced them that the soul dies at the same time as the body and that faith alone can maintain the opposite. However, since the Lateran Council, in its eighth session, held under Pope Leo X,[5] condemns those who make such claims and expressly commands Christian philosophers to refute their arguments and to use their full abilities to demonstrate the truth, I have not hesitated to take on this task as well.

---

3 This quote is from the biblical Book of the Wisdom of Solomon.

4 The Epistle to the Romans is another book of the bible, this time from the New Testament.

5 The Lateran councils were important ecclesiastical councils of the Catholic Church, held at Rome. Descartes refers to the council of 1513 that (among other things) condemned as heretical the philosophical position that the individual soul is not immortal.

Furthermore, I know that several impious people are unwilling to believe that God exists and that the human mind is distinct from the body, for no other reason than, as they allege, no one has been able to prove the truth of these two claims up to now. I in no way agree with these people but, by contrast, believe that almost all [the reasons] great men have brought forward in support of these two questions, once they are sufficiently understood, have the force of demonstrations, and I am convinced that it is virtually impossible to offer any arguments that have not been previously set out by other people. Nevertheless, I consider that there can be no more useful task facing philosophy than to seek out diligently, once and for all, the best of all these arguments and to set them down so accurately and clearly that everyone from now on will accept them as sure proofs. Finally, since I was urgently requested to carry out this work by several people who knew that I had cultivated a certain method for resolving some difficulties in the sciences—not a new method, to be sure, because there is nothing more ancient than the truth, but one which they had often seen me use successfully in other areas—I thought it was my duty to make some sort of attempt at it in this matter.

Whatever I have been able to offer is all contained in this treatise. Not that I have tried to gather together here all the various arguments which one could adduce to serve as demonstrations of the same points, for this did not seem to be worth the effort, except where there was no one proof which was sufficiently certain. Instead, I have described in detail only the first and most important ones in such a way that I now venture to publish them as very certain and very clear demonstrations. Furthermore, I do not believe there is any road open to the human mind by which it is possible ever to come up with better proofs. For the urgency of the subject and the glory of God, to which everything here relates, compel me to speak here of my own work somewhat more freely than I usually do. However, although I believe these arguments are clear and certain, that still does not convince me that they are well suited to everyone's understanding. In geometry there are many works written by Archimedes, Apollonius, Pappus,[6] and others that everyone has accepted on the basis of their clarity and certainty, because, in fact, their arguments are very easy to understand and because none of the stages in their reasoning lacks an accurate and coherent link with what has gone before. Nevertheless, because these works are rather long and demand an assiduously attentive reader, they are understood by relatively few people. And so, although I believe that the clarity and certainty of the demonstrations I use here are equal to, or even better than, those in geometric [proofs], I still fear that there are not many people who 4

6 Archimedes of Syracuse (c. 287 BCE – c. 212 BCE), Apollonius of Perga (c. 262 BCE – c. 190 BCE), Pappus of Alexandria (c. 290 – c. 350).

can grasp them sufficiently, partly because these demonstrations, too, are somewhat long and some depend on others, and, most importantly, because they demand a mind that is entirely free of prejudice and that can easily detach itself from its association with the senses. Besides, we will certainly not find more people in the world well equipped for metaphysics than for geometry. However, there is an additional difference. In geometry, every-5 one is convinced that, as a rule, nothing is written down which has not been clearly proven, and so unskilled readers more frequently make the mistake of approving what is false, because they wish to look as if they understand it, and to avoid appearing to deny what is true. But in philosophy, by contrast, people believe that there is nothing which cannot be disputed on one side or the other, and therefore few of them investigate the truth, while the vast majority, eager to acquire a reputation for genius, boldly assail the most important truths.

Hence, because my arguments, whatever their quality, are exploring philosophical issues, I do not expect to achieve very much with them, unless you assist me with your patronage and protection.[7] Your faculty is held in such high esteem in everyone's mind, and the name Sorbonne carries such great authority, that, with the exception of the sacred councils,[8] there has never been a society in whom people have placed more trust, and not merely in matters of faith, for in human philosophy, too, everyone believes that it is impossible to find anywhere else more perspicuity and solidity, more integrity and wisdom in rendering judgments, than among you. Thus, if you deigned to consider this work sufficiently worth the effort and, first of all, were to correct it—for being aware not only of my humanity but also, above all, of my ignorance, I do not claim that there are no errors in it—and, second, were to add what is lacking, perfect what is insufficiently complete, and illustrate what requires further explanation—if you were to do this on your own or at least give me your advice, so that then I could do it—and, finally, once those arguments in the work proving that God exists and that the mind is different from the body have been established 6 and made as clear as I believe they can be, so that they must be truly accepted as extremely accurate proofs, if you were willing to confirm and publicly endorse them, then, if all this happens, I have no doubt that all the errors which have ever existed concerning these two questions will soon be erased from human minds. For truth itself will quickly see to it that other intellectuals and scholars subscribe to your judgment. And your authority will lead the atheists, who tend to be superficial thinkers rather than people with natural acuity or learning, to set aside their spirit of contradiction and

7 It is disputed whether or not Descartes ever actually received the endorsement of the Sorbonne.

8 Ecumenical councils of top church leaders.

perhaps even to take up arms themselves in support of arguments which they know are considered established truths by all those endowed with real intelligence, in order to avoid appearing as if they do not understand them. And finally all the others will quickly accept the evidence of so many testimonials, and there will no longer be anyone in the world who ventures to call into doubt either that God exists or that the human soul is truly distinct from the body. Given your extraordinary wisdom, you yourselves are able to judge better than anyone else how useful this might be. However, it would not be appropriate for me here to commend further the cause of God and religion to those who have always been the greatest support of the Catholic Church.

PREFACE TO THE READER

7

I have previously touched briefly on questions of God and the human mind in my *Discourse on the Method of Reasoning Correctly and Investigating Truth in the Sciences*, published in French in the year 1637. In that work, to be sure, my purpose was not to treat these questions thoroughly but only to consider them generally and to learn from the judgments of my readers how I ought to address them later on. For these questions seemed to me so important that I judged they should be dealt with more than once. And the road I follow in explaining them is so seldom trodden and so remote from the usual path, that I did not think it would be helpful to explain it at length in French and in a discourse which anyone anywhere might read, in case it encouraged those with weaker minds to believe that they, too, should set out along the same route.

However, in that earlier work I asked all those who came across something they considered objectionable in my writings to do me the favor of advising me what that was. Where my remarks on these questions of God and the soul were concerned, they found nothing worth objecting to, except for two things. These comments I will respond to briefly here, before I undertake a more detailed discussion of these matters.

The first objection is as follows: from the fact that the human mind reflecting on itself does not perceive itself to be anything other than a thinking thing, it does not follow that its nature or essence consists merely in its being a thing that thinks, in the sense that this word *merely* excludes everything else which one might perhaps be able to claim pertains to the nature of the soul. To this objection I reply that in that argument I did not wish to exclude those other attributes from a sequence of thoughts leading to the truth of the matter (which was not really my concern at that time) but only from a sequence following my own perceptions. Thus, what I meant was that I had no distinct awareness of anything which I knew belonged to my essence, other than the fact that I was a thinking thing, or a thing possessing in itself the faculty of thinking. In what follows, however, I will show how from the fact that I know nothing else pertains to my essence, it also follows that there is, in fact, nothing else belonging to it. 8

The second objection is that from the fact that I have within me the idea of something more perfect than myself, it does not follow that the idea itself is more perfect than I am and, even less, that what is represented by this idea exists. However, my answer to this objection is that here an ambiguity lies concealed in the word *idea*. For it can be understood materially, as an operation of my intellect, in which sense it cannot be said that it is more perfect than me, or it can be understood objectively, as the thing represented by that operation. Even if we do not assume that this thing exists outside my intellect, it can still be more perfect than I am because of its essence. How it follows merely from the fact that there is within me the idea of something more perfect than myself that this thing truly exists, I will explain in detail in what follows.

In addition to these objections, I have seen two fairly lengthy works, but 9 they were less concerned to attack my reasoning about these matters than my conclusions, using arguments borrowed from sources common among atheists. But arguments like theirs can have no effect on those who understand my reasoning, and the judgments of many people are so perverse and feeble that they are persuaded by opinions which they have earlier adopted, no matter how false and remote from reason they may be, rather than by a true and firm refutation of these opinions which they hear about later on. And so I am unwilling to respond to those criticisms here, because I wish to avoid having to begin by stating them. I will only make the general point that everything atheists commonly toss out to attack the existence of God always depends upon the fact that we attribute human feelings to God or else attribute so much strength and wisdom to our own minds, that we attempt to determine and understand what God can and ought to do. But we will have no difficulty with this type of objection, provided only that we remember to think of our minds as finite things and of God as beyond our comprehension and infinite.

However, now that I have in one way or another taken a preliminary test of people's opinions, I am here addressing once more the same two questions concerning God and the human mind and at the same time dealing fully with the basic principles of First Philosophy, but not in a way that leads me to expect any praise from the general public or from many readers. In fact, I would even advise people not to read this treatise, unless they are able and willing to meditate seriously with me, to detach their minds from their senses, and at the same time to remove all preconceived notions from their thinking. I know well enough that one finds relatively few readers like that. And as for those who do not take the trouble to understand the 10 order and the connections in my arguments and who are keen to chatter on only about individual conclusions, as many habitually do, such people will not harvest much fruit by reading this treatise. Although they may perhaps find in many parts an occasion to quibble, it will still not be easy for them to make a significant objection or any which merits a reply.

And because I am also not committing myself to satisfying other people in all points immediately and am not arrogant enough to believe that I can foresee everything that will seem difficult to anyone, I will first of all set down in the *Meditations* the very thoughts with whose help I reached, it seems to me, a certain and manifest knowledge of the truth; my aim is to discover whether, using the same arguments which convinced me, I might be able to persuade others. Then, after that, I will reply to the objections of several people of exceptional intelligence and learning to whom I sent these *Meditations* for their perusal, before I submitted them to the printer.[9] For they have made so many and such varied objections, that I venture to hope it will not be easy for other criticisms—at least ones of any importance—to arise in anyone's mind which these people have not already touched on. And so I also urge those who read the *Meditations* not to render judgment on it before they have taken the trouble to read all the objections and my replies to them.

9 These Objections and Replies are not reprinted here.

SYNOPSIS OF THE SIX FOLLOWING MEDITATIONS

12

In the First Meditation I set down the reasons which enable us to place everything in doubt, especially material things, at least as long as we do not have foundations for the sciences different from those we have had up to now. Although at first glance the usefulness of such a widespread doubt is not apparent, it is, in fact, very great, because it frees us from all prejudices, sets down the easiest route by which we can detach our minds from our senses, and finally makes it impossible for us to doubt anymore those things which we later discover to be true.

In the Second Meditation, the mind, using its own unique freedom, assumes that all those things about whose existence it can entertain the least doubt do not exist, and recognizes that during this time it is impossible that it itself does not exist. And that is also extremely useful, because in this way the mind can easily differentiate between those things pertaining to it, that is, to its intellectual nature, and those pertaining to the body. However, since at this point some people may perhaps expect an argument [proving] the immortality of the soul, I think I should warn them that I have tried to 13 avoid writing anything which I could not accurately demonstrate and that, therefore, I was unable to follow any sequence of reasoning other than the one used by geometers. That means I start by setting down everything on which the proposition we are looking into depends, before I reach any conclusions about it. Now, the first and most important prerequisite for understanding the immortality of the soul is to form a conception of the soul that is as clear as possible, one entirely distinct from every conception [we have] of the body. And that I have done in this section. After that, it is essential also for us to know that all those things we understand clearly and distinctly are true in a way which matches precisely how we think of them. This I was unable to prove before the Fourth Meditation. We also need to have a distinct conception of corporeal[10] nature. I deal with that point partly in this Second Meditation and partly in the Fifth and Sixth Meditations, as well. And from these we necessarily infer that all those things we conceive clearly and distinctly as different substances, in the same way we think of the mind and the body, are, in fact, truly different substances,

10 Bodily, physical.

distinct from one another, a conclusion I have drawn in the Sixth Meditation. This conclusion is also confirmed in the same meditation from the fact that we cannot think of the body as anything other than something divisible, and, by contrast, [cannot think of] the mind as anything other than something indivisible. For we cannot conceive of half a mind, in the same way we can with a body, no matter how small. Hence, we realize that their natures are not only different but even, in some respects, opposites. However, I have not pursued the matter any further in this treatise for two reasons: (1) because these points are enough to show that the annihilation of the mind does not follow from the corruption of the body, so we mortals thus ought to entertain hopes of another life; and (2) because the premises on the basis of which we can infer the immortality of the mind depend upon an explanation of all the principles of physics. For (2), first of all, we would have to know that all substances without exception—or those 14 things which, in order to exist, must be created by God—are by their very nature incorruptible and can never cease to exist, unless God, by denying them his concurrence,[11] reduces them to nothing, and then, second, we would have to understand that a body, considered generally, is a substance and thus it, too, never dies. But the human body, to the extent that it differs from other bodies, consists merely of a certain arrangement of parts, with other similar accidental[12] properties; whereas, the human mind is not made up of any accidental properties in this way, but is a pure substance. For even if all the accidental properties of the mind were changed—if, for example, it were to think of different things or have different desires and perceptions, and so on—that would not mean it had turned into a different mind. But the human body becomes something different from the mere fact that the shape of some of its parts has changed. From this it follows that the [human] body does, in fact, perish very easily, but that the mind, thanks to its nature, is immortal.

In the Third Meditation I have set out what seems to me a sufficiently detailed account of my main argument to demonstrate the existence of God. However, in order to lead the minds of the readers as far as possible from the senses, in this section I was unwilling to use any comparisons drawn from corporeal things, and thus many obscurities may still remain. But these, I hope, have later been entirely removed in the replies [I have made] to the objections.[13] For instance, among all the others, there is the

11 The continuous divine action which many Christians think necessary to maintain things in existence.

12 See the section in the introduction to the *Meditations*, "Some Useful Background Information," 1., for explanation of "accidental" here.

13 Descartes refers to the set of objections and replies he published at the end of *Meditations,* not reprinted here.

issue of how the idea of a supremely perfect being, which is present within us, could have so much objective reality that it is impossible for it not to originate from a supremely perfect cause. This is illustrated [in the replies] by the comparison with a wholly perfect machine, the idea of which exists in the mind of some craftsman. For just as the objective ingenuity of this idea must have some cause, that is, the technical skill of this craftsman or of someone else from whom he got the idea, so the idea of God, which is in us, cannot have any cause other than God Himself. 15

In the Fourth Meditation, I establish that all the things which we perceive clearly and distinctly are true, and at the same time I explain what constitutes the nature of falsity; these are things that we have to know both to confirm what has gone before and to understand what still remains. (However, in the meantime I must observe that in this part I do not deal in any way with sin, that is, with errors committed in pursuit of good and evil, but only with those which are relevant to judgments of what is true and false. Nor do I consider matters relevant to our faith or to the conduct of our lives, but merely those speculative truths we can know only with the assistance of our natural light.)

In the Fifth Meditation, I offer a general explanation of corporeal nature and, in addition, also demonstrate the existence of God in a new argument, in which, however, several difficulties may, once again, arise. These I have resolved later in my replies to the objections. And finally, I point out in what sense it is true that the certainty of geometrical demonstrations depends upon a knowledge of God.

Finally, in the Sixth Meditation, I differentiate between the understanding and the imagination and describe the principles of this distinction. I establish that the mind is truly distinct from the body, and I point out how, in spite of that, it is so closely joined to the body that they form, as it were, a single thing. I review all the errors which customarily arise through the senses and explain the ways in which such errors can be avoided. And then finally, I set down all the reasons which enable us to infer the existence of material things. I believe these are useful not because they demonstrate the 16 truth of what they prove—for example, that there truly is a world, that human beings have bodies, and things like that, which no one of sound mind ever seriously doubted—but rather because, when we examine these reasons, we see that they are neither as firm or as evident as those by which we arrive at a knowledge of our own minds and of God, so that the latter are the most certain and most evident of all things which can be known by the human intellect. The proof of this one point was the goal I set out to attain in these *Meditations*. For that reason I am not reviewing here, as they arise [in this treatise], various [other] questions I have dealt with elsewhere.

FIRST MEDITATION

17

Concerning Those Things Which Can Be Called into Doubt

It is now several years since I noticed how from the time of my early youth I had accepted many false claims as true, how everything I had later constructed on top of those [falsehoods] was doubtful, and thus how at some point in my life I needed to tear everything down completely and begin again from the most basic foundations, if I wished to establish something firm and lasting in the sciences. But this seemed an immense undertaking, and I kept waiting until I would be old enough and sufficiently mature to know that no later period of my life would come [in which I was] better equipped to undertake this disciplined enquiry. This reason made me delay the project for so long that I would now be at fault if, by [further] deliberation, I wasted the time still left to carry it out. And so today, when I have conveniently rid my mind of all worries and have managed to find myself secure leisure in solitary withdrawal, I will at 18 last find the time for an earnest and unfettered general demolition of my [former] opinions.

Now, for this task it will not be necessary to show that every opinion I hold is false, something which I might well be incapable of ever carrying out. But reason now convinces me that I should withhold my assent from opinions which are not entirely certain and indubitable, no less than from those which are plainly false; so if I uncover any reason for doubt in each of them, that will be enough to reject them all. For that I will not need to run through them separately, a task that would take forever, because once the foundations are destroyed, whatever is built above them will collapse on its own. Thus, I shall at once assault the very principles upon which all my earlier beliefs rested.

Up to this point, what I have accepted as true I have derived either from the senses or through the senses. However, sometimes I have discovered that these are mistaken, and it is prudent never to place one's entire trust in things which have deceived us even once.

However, although from time to time the senses deceive us about minuscule things or those further away, it could well be that there are still many other matters about which we cannot entertain the slightest doubt, even though we derive [our knowledge] of them from sense experience—for example, the fact that I am now here, seated by the fire, wearing a winter robe, holding this paper in my hands, and so on. And, in fact, how could I deny that these very hands and this whole body are mine, unless 19 perhaps I were to compare myself with certain insane people whose brains are so troubled by the stubborn vapors of black bile[14] that they constantly claim that they are kings, when, in fact, they are very poor, or that they are dressed in purple, when they are nude, or that they have earthenware heads, or are complete pumpkins, or made of glass? But these people are mad, and I myself would appear no less demented if I took something from them and applied it to myself as an example.

A brilliant piece of reasoning! But nevertheless I am a person who sleeps at night and experiences in my dreams all the things these [mad] people do when wide awake, sometimes even less probable ones. How often have I had an experience like this: while sleeping at night, I am convinced that I am here, dressed in a robe and seated by the fire, when, in fact, I am lying between the covers with my clothes off! At the moment, my eyes are certainly wide open and I am looking at this piece of paper, this head which I am moving is not asleep, and I am aware of this hand as I move it consciously and purposefully. None of what happens while I am asleep is so distinct. Yes, of course—but nevertheless I recall other times when I have been deceived by similar thoughts in my sleep. As I reflect on this matter carefully, it becomes completely clear to me that there are no certain indicators which ever enable us to differentiate between being awake and being asleep, and this is astounding; in my confusion I am almost convinced that I may be sleeping.

So then, let us suppose that I am asleep and that these particular details—that my eyes are open, that I am moving my head, that I am stretching out my hand—are not true, and that perhaps I do not even have hands like these or a whole body like this. We must, of course, still concede that the things we see while asleep are like painted images, which could only have been made as representations of real things. And so these general things—these eyes, this head, this hand, and this entire body—at least are not imaginary things but really do exist. For even when painters 20 themselves take great care to form sirens and satyrs with the most unusual shapes, they cannot, in fact, give them natures which are entirely new. Instead, they simply mix up the limbs of various animals or, if they happen to come up with something so new that nothing at all like it has been seen before and thus [what they have made] is completely fictitious and false, nonetheless, at least the colors which make up the picture certainly have

14 One of the four basic bodily fluids then thought to be associated with disease when in imbalance.

to be real. For similar reasons, although these general things—eyes, head, hand, and so on—could also be imaginary, still we are at least forced to concede the reality of certain even simpler and more universal objects, out of which, just as with real colors, all those images of things that are in our thoughts, whether true or false, are formed.

Corporeal nature appears, in general, to belong to this class [of things], as well as its extension,[15] the shape of extended things, their quantity or their size and number, the place where they exist, the time which measures how long they last, and things like that.

Thus, from these facts perhaps we are not reaching an erroneous conclusion [by claiming] that physics, astronomy, medicine, and all the other disciplines which rely upon a consideration of composite objects are indeed doubtful, but that arithmetic, geometry, and the other [sciences] like them, which deal with only the simplest and most general matters and have little concern whether or not they exist in the nature of things, contain something certain and indubitable. For whether I am awake or asleep, two and three always add up to five, a square does not have more than four sides, and it does not seem possible to suspect that such manifest truths could be false.

Nevertheless, a certain opinion has for a long time been fixed in my mind—that there is an all-powerful God who created me and [made me] 21 just as I am. But how do I know He has not arranged things so that there is no earth at all, no sky, no extended thing, no shape, no magnitude, no place, and yet seen to it that all these things appear to me to exist just as they do now? Besides, given that I sometimes judge that other people make mistakes with the things about which they believe they have the most perfect knowledge, might I not in the same way be wrong every time I add two and three together, or count the sides of a square, or do something simpler, if that can be imagined? Perhaps God is unwilling to deceive me in this way, for He is said to be supremely good. But if it is contrary to the goodness of God to have created me in such a way that I am always deceived, it would also seem foreign to His goodness to allow me to be occasionally deceived. The latter claim, however, is not one that I can make.

Perhaps there may really be some people who prefer to deny [the existence of] such a powerful God, rather than to believe that all other things are uncertain. But let us not seek to refute these people, and [let us concede] that everything [I have said] here about God is a fiction. No matter how they assume I reached where I am now, whether by fate, or chance, or a continuous series of events, or in some other way, given that being

15 Something's extension is its spatial magnitude—the volume of space it occupies.

deceived and making mistakes would seem to be something of an imperfection, the less power they attribute to the author of my being, the greater the probability that I will be so imperfect that I will always be deceived. I really do not have a reply to these arguments. Instead, I am finally compelled to admit that there is nothing in the beliefs which I formerly held to be true about which one cannot raise doubts. And this is not a reckless or frivolous opinion, but the product of strong and well-considered reasoning. 22 And therefore, if I desire to discover something certain, in future I should also withhold my assent from those former opinions of mine, no less than [I do] from opinions which are obviously false.

But it is not sufficient to have called attention to this point. I must [also] be careful to remember it. For these habitual opinions constantly recur, and I have made use of them for so long and they are so familiar that they have, as it were, acquired the right to seize hold of my belief and subjugate it, even against my wishes, and I will never give up the habit of deferring to and relying on them, as long as I continue to assume that they are what they truly are: opinions which are to some extent doubtful, as I have already pointed out, but still very probable, so that it is much more reasonable to believe them than to deny them. For that reason, I will not go wrong, in my view, if I deliberately turn my inclination into its complete opposite and deceive myself, [by assuming] for a certain period that these earlier opinions are entirely false and imaginary, until I have, as it were, finally brought the weight of both my [old and my new] prejudices into an equal balance, so that corrupting habits will no longer twist my judgment away from the correct perception of things. For I know that doing this will not, for the time being, lead to danger or error and that it is impossible for me to indulge in excessive distrust, since I am not concerned with actions at this point, but only with knowledge.

Therefore, I will assume that it is not God, who is supremely good and the fountain of truth, but some malicious demon, at once omnipotent and supremely cunning, who has been using all the energy he possesses to deceive me. I will suppose that sky, air, earth, colors, shapes, sounds, and all other external things are nothing but the illusions of my dreams, 23 set by this spirit as traps for my credulity. I will think of myself as if I had no hands, no eyes, no flesh, no blood, nor any senses, and yet as if I still falsely believed I had all these things. I shall continue to concentrate resolutely on this meditation, and if, in doing so, I am, in fact, unable to learn anything true, I will at least do what is in my power and with a resolute mind take care not to agree to what is false or to enable the deceiver to impose anything on me, no matter how powerful and cunning [he may be]. But this task is onerous, and laziness brings me back to my customary way of life. I am like a prisoner who in his sleep may happen to enjoy an imaginary liberty and who, when he later begins to suspect that he is asleep, fears to wake up and willingly cooperates with the pleasing illusions [in order to prolong them]. In this way, I unconsciously slip back into my old opinions and am afraid to wake up, in case from now on I would have to spend the period of challenging wakefulness that follows this peaceful relaxation not in the light, but in the inextricable darkness of the difficulties I have just raised.

SECOND MEDITATION

Concerning the Nature of the Human Mind and the Fact that It Is Easier to Know than the Body

Yesterday's meditation threw me into so many doubts that I can no longer forget them or even see how they might be resolved. Just as if I had suddenly fallen into a deep eddying current, I am hurled into such 24 confusion that I am unable to set my feet on the bottom or swim to the surface. However, I will struggle along and try once again [to follow] the same path I started on yesterday—that is, I will reject everything which admits of the slightest doubt, just as if I had discovered it was completely false, and I will proceed further in this way, until I find something certain, or at least, if I do nothing else, until I know for certain that there is nothing certain. In order to shift the entire earth from its location, Archimedes asked for nothing but a fixed and immovable point. So I, too, ought to hope for great things if I can discover something, no matter how small, which is certain and immovable.

Therefore, I assume that everything I see is false. I believe that none of those things my lying memory represents has ever existed, that I have no senses at all, and that body, shape, extension, motion, and location are chimeras.[16] What, then, will be true? Perhaps this one thing: there is nothing certain.

But how do I know that there exists nothing other than the items I just listed, about which one could not entertain the slightest momentary doubt? Is there not some God, by whatever name I call him, who places these very thoughts inside me? But why would I think this, since I myself could perhaps have produced them? So am I then not at least something? But I have already denied that I have senses and a body. Still, I am puzzled, for 25 what follows from this? Am I so bound up with my body and my senses that I cannot exist without them? But I have convinced myself that there

16 In Greek mythology, a female fire-breathing monster with a lion's head, a goat's body, and a serpent's tail; more generally, an absurd or horrible idea or wild fancy.

is nothing at all in the universe—no sky, no earth, no minds, no bodies. So then, is it the case that I, too, do not exist? No, not at all: if I persuaded myself of something, then I certainly existed. But there is some kind of deceiver, supremely powerful and supremely cunning, who is constantly and intentionally deceiving me. But then, if he is deceiving me, there again is no doubt that I exist—for that very reason. Let him trick me as much as he can, he will never succeed in making me nothing, as long as I am aware that I am something. And so, after thinking all these things through in great detail, I must finally settle on this proposition: the statement *I am, I exist* is necessarily true every time I say it or conceive of it in my mind.

But I do not yet understand enough about what this *I* is, which now necessarily exists. Thus, I must be careful I do not perhaps unconsciously substitute something else in place of this *I* and in that way make a mistake even here, in the conception which I assert is the most certain and most evident of all. For that reason, I will now reconsider what I once believed myself to be, before I fell into this [present] way of thinking. Then I will remove from that whatever could, in the slightest way, be weakened by the reasoning I have [just] brought to bear, so that, in doing this, by the end I will be left only with what is absolutely certain and immovable.

What then did I believe I was before? Naturally, I thought I was a human being. But what is a human being? Shall I say a *rational animal*? No. For then I would have to ask what an *animal* is and what *rational* means, and thus from a single question I would fall into several greater difficulties. And at the moment I do not have so much leisure time that I wish to squander it with subtleties of this sort. Instead I would prefer here to attend 26 to what used to come into my mind quite naturally and spontaneously in earlier days every time I thought about what I was. The first thought, of course, was that I had a face, hands, arms, and this entire mechanism of limbs, the kind one sees on a corpse, and this I designated by the name *body*. Then it occurred to me that I ate and drank, walked, felt, and thought. These actions I assigned to the *soul*. But I did not reflect on what this *soul* might be, or else I imagined it as some kind of attenuated substance, like wind, or fire, or aether,[17] spread all through my denser parts. However, I had no doubts at all about my body—I thought I had a clear knowledge of its nature. Perhaps if I had attempted to describe it using the mental conception I used to hold, I would have explained it as follows: By a *body* I understand everything that is appropriately bound together in a certain form and confined to a place; it fills a certain space in such a way as to

17 Aether is the fifth element of medieval alchemy, and the idea has its origins in the classical Greek notion of the pure atmosphere beyond the sky in which the gods were thought to live, and which they breathed, analogous to (but different from) the air of the terrestrial atmosphere.

exclude from that space every other body; it can be perceived by touch, sight, hearing, taste, or smell, and can also be moved in various ways, not, indeed, by itself, but by something else which makes contact with it. For I judged that possessing the power of self-movement, like the ability to perceive things or to think, did not pertain at all to the nature of body. Quite the opposite in fact, so that when I found out that faculties rather similar to these were present in certain bodies, I was astonished.

But what [am I] now, when I assume that there is some extremely powerful and, if I may be permitted to speak like this, malevolent and deceiving being who is deliberately using all his power to trick me? Can I affirm that I possess even the least of all those things which I have just described 27 as pertaining to the nature of body? I direct my attention [to this], think [about it], and turn [the question] over in my mind. Nothing comes to me. It is tedious and useless to go over the same things once again. What, then, of those things I used to attribute to the soul, like eating, drinking, or walking? But given that now I do not possess a body, these are nothing but imaginary figments. What about sense perception? This, too, surely does not occur without the body. And in sleep I have apparently sensed many objects which I later noticed I had not [truly] perceived. What about thinking? Here I discover something: thinking does exist. This is the only thing which cannot be detached from me. *I am, I exist*—that is certain. But for how long? Surely for as long as I am thinking. For it could perhaps be the case that, if I were to abandon thinking altogether, then in that moment I would completely cease to be. At this point I am not agreeing to anything except what is necessarily true. Therefore, strictly speaking, I am merely a thinking thing, that is, a mind or spirit, or understanding, or reason—words whose significance I did not realize before. However, I am something real, and I truly exist. But what kind of thing? As I have said, a thing that thinks.

And what else besides? I will let my imagination roam. I am not that interconnection of limbs we call a human body. Nor am I even some attenuated air which filters through those limbs—wind, or fire, or vapor, or breath, or anything I picture to myself. For I have assumed those things were nothing. Let this assumption hold. Nonetheless, I am still something. Perhaps it could be the case that these very things which I assume are nothing, because they are unknown to me, are truly no different from that *I* which I do recognize. I am not sure, and I will not dispute this point right now. I can render judgment only on those things which are known to me: I know that I exist. I am asking what this *I* is—the thing I know. It is very certain that knowledge of this *I*, precisely defined like this, does not de- 28 pend on things whose existence I as yet know nothing about and therefore on any of those things I conjure up in my imagination. And this phrase *conjure up* warns me of my mistake, for I would truly be conjuring something up if I imagined myself to be something, since imagining is nothing other than contemplating the form or the image of a physical thing. But now I know for certain that I exist and, at the same time, that it is possible for all those images and, in general, whatever relates to the nature of body to be nothing but dreams [or chimeras]. Having noticed this, it seems no less foolish for me to say "I will let my imagination work, so that I may recognize more clearly what I am" than if I were to state, "Now I am indeed awake, and I see some truth, but because I do yet not see it with sufficient clarity, I will quite deliberately go to sleep, so that in my dreams I will get a truer and more distinct picture of it." Therefore, I realize that none of those things which I can understand with the aid of my imagination is pertinent to this idea I possess about myself and that I must be extremely careful to summon my mind back from such things, so that it may perceive its own nature with the utmost clarity, on its own.

But what then am I? A thinking thing. What is this? It is surely something that doubts, understands, affirms, denies, is willing, is unwilling, and also imagines and perceives.

This is certainly not an insubstantial list, if all [these] things belong to me. But why should they not? Surely I am the same *I* who now doubts almost everything, yet understands some things, who affirms that this one thing is true, denies all the rest, desires to know more, does not wish to be deceived, imagines many things, even against its will, and also notices many things which seem to come from the senses? Even if I am always asleep and even if the one who created me is also doing all he can to deceive me, what is there among all these things which is not just as true as 29 the fact that I exist? Is there something there that I could say is separate from me? For it is so evident that I am the one who doubts, understands, and wills, that I cannot think of anything which might explain the matter more clearly. But obviously it is the same *I* that imagines, for although it may well be the case, as I have earlier assumed, that nothing I directly imagine is true, nevertheless, the power of imagining really exists and forms part of my thinking. Finally, it is the same *I* that feels, or notices corporeal things, apparently through the senses: for example, I now see light, hear noise, and feel heat. But these are false, for I am asleep. Still, I certainly seem to see, hear, and grow warm—and this cannot be false. Strictly speaking, this is what in me is called sense perception and, taken in this precise meaning, it is nothing other than thinking.

From these thoughts, I begin to understand somewhat better what I am. However, it still appears that I cannot prevent myself from thinking that corporeal things, whose images are formed by thought and which the senses themselves investigate, are much more distinctly known than that obscure part of me, the *I*, which is not something I can imagine, even though it is really strange that I have a clearer sense of those things whose existence I know is doubtful, unknown, and alien to me than I do of something which is true and known, in a word, of my own self. But I realize what the trouble is. My mind loves to wander and is not yet allowing itself to be confined within the limits of the truth. All right, then, let us at this point for once give it completely free rein, so that a little later on, when 30 the time comes to pull back, it will consent to be controlled more easily.

Let us consider those things we commonly believe we understand most distinctly of all, that is, the bodies we touch and see—not, indeed, bodies in general, for those general perceptions tend to be somewhat more confusing, but rather one body in particular. For example, let us take this [piece of] beeswax. It was collected from the hive very recently and has not yet lost all the sweetness of its honey. It [still] retains some of the scent of the flowers from which it was gathered. Its color, shape, and size are evident. It is hard, cold, and easy to handle. If you strike it with your finger, it will give off a sound. In short, everything we require to be able to recognize a body as distinctly as possible appears to be present. But watch. While I am speaking, I bring the wax over to the fire. What is left of its taste is removed, its smell disappears, its color changes, its shape is destroyed, its size increases, it turns to liquid, and it gets hot. I can hardly touch it. And now, if you strike it, it emits no sound. After [these changes], is what remains the same wax? We must concede that it is. No one denies this; no one thinks otherwise. What then was in [this piece of wax] that I understood so distinctly? Certainly nothing I apprehended with my senses, since all [those things] associated with taste, odor, vision, touch, and sound have now changed. [But] the wax remains.

Perhaps what I now think is as follows: the wax itself was not really that sweetness of honey, that fragrance of flowers, that white color, or that shape and sound, but a body which a little earlier was perceptible to me in those forms, but which is now [perceptible] in different ones. But what exactly is it that I am imagining in this way? Let us consider that point 31 and, by removing those things which do not belong to the wax, see what is left over. It is clear that nothing [remains], other than something extended, flexible, and changeable. But what, in fact, do *flexible* and *changeable* mean? Do these words mean that I imagine that this wax can change from a round shape to a square one or from [something square] to something triangular? No, that is not it at all. For I understand that the wax has the capacity for innumerable changes of this kind, and yet I am not able to run through these innumerable changes by using my imagination. Therefore, this conception [I have of the wax] is not produced by the faculty of imagination. What about extension? Is not the extension of the wax also unknown? For it becomes greater when the wax melts, greater [still] when it boils, and once again [even] greater, if the heat is increased. And I would not be judging correctly what wax is if I did not believe that it could also be extended in various other ways, more than I could ever grasp in my imagination. Therefore, I am forced to admit that my imagination has no idea at all what this wax is and that I perceive it only with my mind. I am talking about this [piece of] wax in particular, for the point is even clearer about wax in general. But what is this wax which can be perceived only by the mind? It must be the same as the wax I see, touch, and imagine—in short, the same wax I thought it was from the beginning. But we should note that the perception of it is not a matter of sight, or touch, or imagination, and never was, even though that seemed to be the case earlier, but simply of mental inspection, which could be either imperfect and confused as it was before, or clear and distinct as it is now, depending on the lesser or greater degree of attention I bring to bear on those things out of which the wax is composed.

However, now I am amazed at how my mind is [weak and] prone to error. For although I am considering these things silently within myself, 32 without speaking aloud, I still get stuck on the words themselves and am almost deceived by the very nature of the way we speak. For if the wax is there [in front of us], we say that we see the wax itself, not that we judge it to be there from the color or shape. From that I could immediately conclude that I recognized the wax thanks to the vision in my eyes, and not simply by mental inspection. But by analogy, suppose I happen to glance out of the window at people crossing the street; in normal speech I also say I see the people themselves, just as I do with the wax. But what am I really seeing other than hats and coats, which could be concealing automatons[18] underneath? However, I judge that they are people. And thus what I thought I was seeing with my eyes I understand only with my faculty of judgment, which is in my mind.

But someone who wishes [to elevate] his knowledge above the common level should be ashamed to have based his doubts in the forms of speech which ordinary people use, and so we should move on to consider next whether my perception of what wax is was more perfect and more evident when I first perceived it and believed I knew it by my external senses, or at least by my so-called *common sense*,[19] in other words, by the power of imagination, or whether it is more perfect now, after I have investigated more carefully both what wax is and how it can be known. To entertain doubts about this matter would certainly be silly. For in my first perception of the wax what was distinct? What did I notice there that any animal might not be capable of capturing? But when I distinguish the wax from

18 Mechanical person-imitations; robots.

19 This is the supposed mental faculty which unites the data from the five external senses—sight, smell, sound, touch, and taste—into a single sensory experience. The notion goes back to Aristotle, and is different from what we call "common sense" today.

its external forms and look at it as something naked, as if I had stripped off its clothing, even though there could still be some error in my judgment, it is certain that I could not perceive it in this way without a human mind.

But what am I to say about this mind itself, in other words, about my-33 self? For up to this point I am not admitting there is anything in me except mind. What, I say, is the *I* that seems to perceive this wax so distinctly? Do I not know myself not only much more truly and certainly, but also much more distinctly and clearly than I know the wax? For if I judge that the wax exists from the fact that I see it, then from the very fact that I see the wax it certainly follows much more clearly that I myself also exist. For it could be that what I see is not really wax. It could be the case that I do not have eyes at all with which to see anything. But when I see or think I see (at the moment I am not differentiating between these two), it is completely impossible that I, the one doing the thinking, am not something. For similar reasons, if I judge that the wax exists from the fact that I am touching it, the same conclusion follows once again, namely, that I exist. The result is clearly the same if [my judgment rests] on the fact that I imagine the wax or on any other reason at all. But these observations I have made about the wax can be applied to all other things located outside of me. Furthermore, if my perception of the wax seemed more distinct after it was drawn to my attention, not merely by sight or touch, but by several [other] causes, I must concede that I now understand myself much more distinctly, since all of those same reasons capable of assisting my perception either of the wax or of any other body whatsoever are even better proofs of the nature of my mind! However, over and above this, there are so many other things in the mind itself which can provide a more distinct conception of its [nature] that it hardly seems worthwhile to review those features of corporeal things which might contribute to it.

And behold—I have all on my own finally returned to the place where 34 I wanted to be. For since I am now aware that bodies themselves are not properly perceived by the senses or by the faculty of imagination, but only by the intellect, and are not perceived because they are touched or seen, but only because they are understood, I realize this obvious point: there is nothing I can perceive more easily or more clearly than my own mind. But because it is impossible to rid oneself so quickly of an opinion one has long been accustomed to hold, I would like to pause here, in order to impress this new knowledge more deeply on my memory with a prolonged meditation.

THIRD MEDITATION

Concerning God and the Fact that He Exists

Now I will close my eyes, stop up my ears, and withdraw all my senses. I will even blot out from my thinking all images of corporeal things, or else, since this is hardly possible, I will dismiss them as empty and false images of nothing at all, and by talking only to myself and looking more deeply within, I will attempt, little by little, to acquire a greater knowledge of and more familiarity with myself. I am a thinking thing—in other words, something that doubts, affirms, denies, knows a few things, is ignorant of many things, wills, refuses, and also imagines and feels. For, as I have pointed out earlier, although those things which I sense or imagine outside of myself are perhaps nothing, nevertheless, I am certain that the thought processes I call sense experience and imagination, given that they 35 are only certain modes of thinking, do exist within me.

In these few words, I have reviewed everything I truly know, or at least [everything] that, up to this point, I was aware I knew. Now I will look around more diligently, in case there are perhaps other things in me that I have not yet considered. I am certain that I am a thinking thing. But if that is the case, do I not then also know what is required for me to be certain about something? There is, to be sure, nothing in this first knowledge other than a certain clear and distinct perception of what I am affirming, and obviously this would not be enough for me to be certain about the truth of the matter, if it could ever happen that something I perceived just as clearly and distinctly was false. And now it seems to me that now I can propose the following general rule: all those things I perceive very clearly and very distinctly are true.

However, before now I have accepted as totally certain and evident many things that I have later discovered to be doubtful. What, then, were these things? [They were], of course, the earth, the sky, the stars, and all the other things I used to grasp with my senses. But what did I clearly perceive in them? Obviously I was observing in my mind ideas or thoughts of such things. And even now I do not deny that those ideas exist within me. However, there was something else which I held to be true and which, because I was in the habit of believing it, I also thought I perceived clearly, although I really was not perceiving it at all, namely, that certain things existed outside of me from which those ideas proceeded and which were like them in every way. And here was where I went wrong, or, if anyway I was judging truthfully, that certainly was not the result of the strength of my perception.

What [then was] true? When I was thinking about something very sim-
36 ple and easy in arithmetic or geometry—for example, that two and three added together make five, and things of that sort—was I not recognizing these with sufficient clarity at least to affirm that they were true? Later on, to be sure, I did judge that such things could be doubted, but the only reason I did so was that it crossed my mind that some God could perhaps have placed within me a certain kind of nature, so that I deceived myself even about those things which appeared most obvious. And every time this preconceived opinion about the supreme power of God occurs to me, I cannot but confess that if He wished, it would be easy for Him to see to it that I go astray, even in those matters which I think I see as clearly as possible with my mind's eye. But whenever I turn my attention to those very things which I think I perceive with great clarity, I am so completely persuaded by them, that I spontaneously burst out with the following words: Let whoever can deceive me, do so; he will still never succeed in making me nothing, not while I think I am something, or in making it true someday that I never existed, since it is true that I exist now, or perhaps even in making two and three, when added together, more or less than five, or anything like that, in which I clearly recognize a manifest contradiction. And since I have no reason to think that some God exists who is a deceiver and since, up to this point, I do not know enough to state whether there is a God at all, it is clear that the reason for any doubt which rests on this supposition alone is very tenuous and, if I may say so, metaphysical. However, to remove even that doubt, as soon as the occasion presents itself, I ought to examine whether God exists and, if He does, whether He can be a deceiver. For as long as this point remains obscure, it seems to me that I can never be completely certain about anything else.

But now an orderly arrangement would seem to require that I first divide all of my thoughts into certain kinds and look into which of these
37 [kinds], strictly speaking, contain truth or error. Some of my thoughts are, so to speak, images of things, and for these alone the name *idea* is appropriate, for example, when I think of a man, or a chimera, or the sky, or an angel, or God. But other thoughts, in addition to these, possess certain other forms. For example, when I will, when I fear, when I affirm, and when I deny, I always apprehend something as the object of my thinking, but in my thought I also grasp something more than the representation of that thing. In this [group of thoughts], some are called volitions[20] or feelings, and others judgments.

20 Acts of decision-making.

Now, where ideas are concerned, if I consider these only in and of themselves, not considering whether they refer to anything else, they cannot, strictly speaking, be false. For whether I imagine a goat or a chimera, it is no less true that I imagine one than it is that I imagine the other. And we also need have no fear of error in willing or in feeling, for although I can desire something evil or even things which have never existed, that still does not make the fact that I desire them untrue. And thus, all that remains are judgments, in which I must take care not to be deceived. But the most important and most frequent error I can discover in judgments consists of the fact that I judge the ideas within me are similar to or conform to certain things located outside myself. For obviously, if I considered ideas themselves only as certain modes of my thinking, without considering their reference to anything else, they would hardly furnish me any material for making a mistake.

Of these ideas, some, it seems to me, are innate,[21] others come from outside, and still others I have myself made up. For the fact that I understand what a thing is, what truth is, and what thinking is I seem to possess from no source other than my own nature. But if I now hear a noise, see the sun, or feel heat, I have up to now judged that [these sensations] come from certain things placed outside of me. And, finally, sirens, hippogriffs,[22] and such like are things I myself dream up. But perhaps I could also believe that all [these ideas] come from outside, or else are all innate, or else are all made up, for I have not yet clearly perceived their true origin. 38

However, the most important point I have to explore here concerns those ideas which I think of as being derived from objects existing outside me: What reason leads me to suppose that these ideas are similar to those objects? It certainly seems that I am taught to think this way by nature. Furthermore, I know by experience that these [ideas] do not depend on my will and therefore not on me myself, for they often present themselves to me even against my will. For example, whether I will it or not, I now feel heat, and thus I believe that the feeling or the idea of heat reaches me from some object apart from me, that is, from [the heat] of the fire I am sitting beside. And nothing is more obvious than my judgment that this object is sending its own likeness into me rather than something else.

I will now see whether these reasons are sufficiently strong. When I say here that I have been taught to think this way by nature, I understand only that I have been carried by a certain spontaneous impulse to believe it, not that some natural light has revealed its truth to me. There is an important difference between these two things. For whatever natural light reveals to

---

21 Inborn—an idea that is already inside me.

22 In Greek mythology, sirens are half woman, half bird; hippogriffs are combinations of horse and griffin (which is part eagle, part lion).

me—for example, that from the fact that I am doubting it follows that I exist, and things like that—cannot admit of any possible doubt, because there cannot be another faculty [in me] as trustworthy as natural light, one which could teach me that the ideas [derived from natural light] are not 39 true. But where natural impulses are concerned, in the past, when there was an issue of choosing the good thing to do, I often judged that such impulses were pushing me in the direction of something worse, and I do not see why I should place more trust in them in any other matters.

Moreover, although those ideas do not depend on my will, it is not therefore the case that they must come from objects located outside of me. For just as those impulses I have been talking about above are within me and yet seem to be different from my will, so perhaps there is also some other faculty in me, one I do not yet understand sufficiently, which produces those ideas, in the same way they have always appeared to be formed in me up to now while I sleep, without the help of any external objects [which they represent].

Finally, even if these ideas did come from things different from me, it does not therefore follow that they have to be like those things. Quite the contrary, for in numerous cases I seem to have often observed a great difference [between the object and the idea]. So, for example, I find in my mind two different ideas of the sun. One, which is apparently derived from the senses and should certainly be included among what I consider ideas coming from outside, makes the sun appear very small to me. However, the other, which is derived from astronomical reasoning, that is, elicited by certain notions innate in me or else produced by me in some other manner, makes the sun appear many times larger than the earth. Clearly, these two [ideas] cannot both resemble the sun which exists outside of me, and reason convinces [me] that the one which seems to have emanated most immediately from the sun itself is the least like it.

40 All these points offer me sufficient proof that previously, when I believed that certain things existed apart from me that conveyed ideas or images of themselves, whether by my organs of sense or by some other means, my judgment was not based on anything certain but only on some blind impulse.

However, it crosses my mind that there is still another way of exploring whether certain things of which I have ideas within me exist outside of me. To the extent that those ideas are [considered] merely certain ways of thinking, of course, I do not recognize any inequality among them, and they all appear to proceed from me in the same way. But to the extent that one idea represents one thing, while another idea represents something else, it is clear that they are very different from each other. For undoubtedly those that represent substances to me and contain in themselves more objective reality, so to speak, are something more than those that simply represent modes or accidents.[23] And, once again, that idea thanks to which I am aware of a supreme God—eternal, infinite, omniscient, omnipotent, the Creator of all things that exist outside of Him—certainly has more objective reality in it than those ideas through which finite substances are represented.

Now, it is surely evident by natural light that there must be at least as much [reality] in the efficient and total cause as there is in the effect of this cause. For from where, I would like to know, can the effect receive its reality if not from its cause? And how can the cause provide this reality to the effect, unless the cause also possesses it? But from this it follows that something cannot be made from nothing and also that what is more perfect, that is, contains more reality in itself, cannot be produced from what is less perfect. This is obviously true not only of those effects whose reality 41 is [what the philosophers call] actual or formal, but also of those ideas in which we consider only [what they call] objective reality. For example, some stone which has not previously existed cannot now begin to exist, unless it is produced by something which has in it, either formally or eminently, everything that goes into the stone,[24] and heat cannot be brought into an object which was not warm previously, except by something which is of an order at least as perfect as heat, and so on with all the other examples. But beyond this, even the idea of heat or of the stone cannot exist within me, unless it is placed in me by some cause containing at least as much reality as I understand to be in the heat or in the stone. For although that cause does not transfer anything of its own reality, either actual or formal, into my idea, one should not therefore assume that [this cause] must be less real. Instead, [we should consider] that the nature of the idea itself is such that it requires from itself no formal reality other than what it derives from my own thinking, of which it is a mode [that is, a way or style of thinking]. But for the idea to possess this objective reality rather than another, it must surely obtain it from some cause in which there is at least as much formal reality as the objective reality contained in the idea itself. For if we assume that something can be discovered in the idea which was not present in its cause, then it must have obtained this from nothing. But no matter how imperfect the mode of being may be by which a thing is objectively present in the understanding through its idea, that mode is certainly not nothing, and therefore [this idea] cannot come from nothing.

23 See the introduction to the *Meditations* for background information on "substance," "accident," "objective reality," and "formal reality."

24 That is, it has either the same properties as the stone (e.g., a certain hardness) or possesses even more perfect or pronounced versions of those properties (e.g., perfect hardness). An effect is "eminently" in a cause when the cause is more perfect than the effect.

And although the reality which I am considering in my ideas is only 42 objective, I must not imagine that it is unnecessary for the same reality to exist formally in the causes of those ideas, that it is sufficient if [the reality] in them is objective as well. For just as that mode of existing objectively belongs to ideas by their very nature, so the mode of existing formally belongs to the causes of [these] ideas, at least to the first and most important causes, by their nature. And although it may well be possible for one idea to be born from another, still this regress cannot continue on *ad infinitum*,[25] for we must finally come to some first [idea], whose cause is, as it were, the archetype [or original idea], which formally contains the entire reality that exists only objectively in the idea. And thus natural light makes it clear to me that ideas exist within me as certain images that can, in fact, easily fall short of the perfection of the things from which they were derived but that cannot contain anything greater or more perfect than those things do.

And the more time and care I take examining these things, the more clearly and distinctly I recognize their truth. But what am I finally to conclude from them? It is clear that if the objective reality of any of my ideas is so great that I am certain that the same reality is not in me either formally or eminently and that therefore I myself cannot be the cause of that idea, it necessarily follows that I am not alone in the world but that some other thing also exists which is the cause of that idea. But if I do not find any such idea within me, then I will obviously have no argument that confirms for me the existence of anything beyond myself. For I have been searching very diligently and have not been able to find any other argument up to now.

But of these ideas of mine, apart from the one which reveals my own 43 self to me, about which there can be no difficulty, there is another [that represents] God [to me], and there are others which represent corporeal and inanimate things, as well as others representing angels, animals, and finally other men who resemble me.

As far as concerns those ideas which display other human beings or animals or angels, I understand readily enough that I could have put these together from ideas I have of myself, of corporeal things, and of God, even though there might be no people apart from me, or animals or angels in the world.

Where the ideas of corporeal things are concerned, I see nothing in them so great that it seems as if it could not have originated within me. For if I inspect these ideas thoroughly and examine them individually in the same way I did yesterday with the idea of the wax, I notice that there are only a very few things I perceive in them clearly and distinctly—for example, magnitude or extension in length, breadth, and depth; shape,

25 Forever (to infinity).

which emerges from the limits of that extension; position, which different forms derive from their relation to each other; and motion or a change of location. To these one can add substance, duration, and number. However, with the other things, like light, colors, sounds, odors, tastes, heat, cold, and other tactile qualities, my thoughts of them involve so much confusion and obscurity, that I still do not know whether they are true or false—in other words, whether the ideas I have of these [qualities] are ideas of things or of non-things. For although I observed a little earlier that falsehood (or, strictly speaking, formal falsehood) could occur only in judgments, nonetheless there is, in fact, a certain other material falsehood in ideas, when they represent a non-thing as if it were a thing. Thus, for example, ideas which I have of heat and cold are so unclear and indistinct that I am 44 not able to learn from them whether cold is merely a lack of heat, or heat a lack of cold, or whether both of these are real qualities, or whether neither [of them is]. And because there can be no ideas which are not, as it were, ideas of things, if it is indeed true that cold is nothing other than a lack of heat, the idea which represents cold to me as if it were something positive and real will not improperly be called false, and that will also hold for all other ideas [like this].

To such ideas I obviously do not have to assign any author other than myself, for, if they are, in fact, false, that is, if they represent things which do not exist, my natural light informs me that they proceed from nothing—in other words, that they are in me only because there is something lacking in my nature, which is not wholly perfect. If, on the other hand, they are true, given that the reality they present to me is so slight that I cannot distinguish the object from something which does not exist, then I do not see why I could not have come up with them myself.

As for those details which are clear and distinct in my ideas of corporeal things, some of them, it seems to me, I surely could have borrowed from the idea of myself, namely, substance, duration, number, and other things like that. I conceive of myself as a thinking and non-extended thing, but of the stone as an extended thing which does not think; so there is a great difference between the two; but nevertheless, I think of both as *substance*, something equipped to exist on its own. In the same way, when I perceive that I now exist and also remember that I have existed for some time earlier, and when I have various thoughts whose number I recognize, I acquire 45 ideas of *duration* and *number*, which I can then transfer to any other things I choose. As for all the other qualities from which I put together my ideas of corporeal things, that is, extension, shape, location, and motion, they are, it is true, not formally contained in me, since I am nothing other than a thinking thing, but because they are merely certain modes of a substance and I, too, am a substance, it seems that they could be contained in me eminently.

And so the only thing remaining is the idea of God. I must consider whether there is anything in this idea for which I myself could not have been the origin. By the name *God* I understand a certain infinite, [eternal, immutable,] independent, supremely intelligent, and supremely powerful substance by which I myself was created, along with everything else that exists (if, [in fact], anything else does exist). All of these [properties] are clearly [so great] that the more diligently I focus on them, the less it seems that I could have brought them into being by myself alone. And thus, from what I have said earlier, I logically have to conclude that God necessarily exists.

For although the idea of a substance is, indeed, in me—because I am a substance—that still does not mean [that I possess] the idea of an infinite substance, since I am finite, unless it originates in some other substance which is truly infinite.

And I should not think that my perception of the infinite comes, not from a true idea, but merely from a negation of the finite, in [the same] way I perceive rest and darkness by a negation of motion and light. For, on the contrary, I understand clearly that there is more reality in an infinite substance than in a finite one and that therefore my perception of the infinite is somehow in me before my perception of the finite—in other words, my perception of God comes before my perception of myself. 46 For how would I know that I am doubting or desiring, or, in other words, that something is lacking in me and that I am not entirely perfect, unless some idea of a perfect being was in me and I recognized my defects by a comparison?

And one cannot claim that this idea of God might well be materially false and thus could have come from nothing, the way I observed a little earlier with the ideas of heat and cold and things like that. Quite the reverse: for [this idea] is extremely clear and distinct and contains more objective reality than any other, and thus no idea will be found which is more inherently true and in which there is less suspicion of falsehood. This idea, I say, of a supremely perfect and infinite being is utterly true, for although it may well be possible to imagine that such a being does not exist, it is still impossible to imagine that the idea of Him does not reveal anything real to me, in the way I talked above about the idea of cold. This idea of a perfect Being is also entirely clear and distinct, for whatever I see clearly and distinctly which is real and true and which introduces some perfection is totally contained within [this idea]. The fact that I cannot comprehend the infinite or that there are innumerable other things in God that I do not understand or even perhaps have any way of contacting in my thoughts—all this is irrelevant. For something finite, like myself, cannot comprehend the nature of the infinite, and it is sufficient that I understand this very point and judge that all things which I perceive clearly and which I know convey some perfection, as well as innumerable others perhaps which I know nothing about, are in God, either formally or eminently, so that the idea I have of Him is the truest, clearest, and most distinct of all the ideas within me.

But perhaps I am something more than I myself understand, and all those perfections which I attribute to God are potentially in me somehow, even though they are not yet evident and are not manifesting themselves 47 in action. For I already know by experience that my knowledge is gradually increasing, and I do not see anything which could prevent it from increasing more and more to infinity. Nor do I even know of any reasons why, with my knowledge augmented in this way, I could not, with its help, acquire all the other perfections of God or, finally, why, if the power [to acquire] those perfections is already in me, it would not be sufficient to produce the idea of those perfections.

And yet none of these things is possible. For, in the first place, although it is true that my knowledge is gradually increasing and that there are potentially many things within me which have not yet been realized, still none of these is relevant to the idea of God, in which, of course, nothing at all exists potentially. For the very fact that my knowledge is increasing little by little is the most certain argument for its imperfection. Beyond that, even if my knowledge is always growing more and more, nonetheless, that does not convince me that it will ever be truly infinite, since it can never reach a stage where it is not capable of increasing any further. But I judge that God is actually infinite, so that nothing can possibly be added to His perfection. And lastly, I perceive that the objective existence of an idea cannot be produced from a being that is merely potential, which, strictly speaking, is nothing, but only from something which actually or formally exists.

Obviously everything in all these thoughts is evident to the natural light in anyone who reflects carefully [on the matter]. But when I pay less attention and when images of sensible[26] things obscure the vision in my mind, I do not so readily remember why the idea of a being more perfect than myself must necessarily proceed from some entity that is truly more 48 perfect than me. Therefore, I would like to enquire further whether I, who possess this idea [of God], could exist if such a being did not exist.

If that were the case, then from whom would I derive my existence? Clearly from myself or from my parents or from some other source less perfect than God. For we cannot think of or imagine anything more perfect than God or even anything equally perfect.

However, if I originated from myself, then I would not doubt or hope, and I would lack nothing at all, for I would have given myself all the perfections of which I have any idea within me, and thus I myself would

26 I.e., things that can be perceived with the physical senses.

be God. I must not assume that those things which I lack might be more difficult to acquire than those now within me. On the contrary, it clearly would have been much more difficult for me—that is, a thinking thing or substance—to emerge from nothing than to acquire a knowledge of the many things about which I am ignorant, for knowing such things is merely an accident of that thinking substance. And surely if I had obtained from myself that greater perfection [of being the author of my own existence], then I could hardly have denied myself the perfections which are easier to acquire, or, indeed, any of those I perceive contained in the idea of God, since, it seems to me, none of them is more difficult to produce. But if there were some perfections more difficult to acquire, they would certainly appear more difficult to me, too, if, indeed, everything else I possessed was derived from myself, because from them I would learn by experience that my power was limited.

And I will not escape the force of these arguments by assuming that I might perhaps have always been the way I am now, as if it followed from that assumption that I would not have to seek out any author for my own 49 existence. For since the entire period of my life can be divided into innumerable parts each one of which is in no way dependent on the others, therefore, just because I existed a little while ago, it does not follow that I must exist now, unless at this very moment some cause is, at it were, creating me once again—in other words, preserving me. For it is clear to anyone who directs attention to the nature of time that, in order for the existence of anything at all to be preserved in each particular moment it lasts, that thing surely needs the same force and action which would be necessary to create it anew if it did not yet exist. Thus, one of the things natural light reveals is that preservation and creation are different only in the ways we think of them.

Consequently, I now ought to ask myself whether I have any power which enables me to bring it about that I, who am now existing, will also exist a little later on, for since I am nothing other than a thinking thing—or at least since my precise concern at the moment is only with that part of me which is a thinking thing—if such a power is in me, I would undoubtedly be conscious of it. But I experience nothing [of that sort], and from this fact alone I recognize with the utmost clarity that I depend upon some being different from myself.

But perhaps that being is not God, and I have been produced by my parents or by some other causes less perfect than God. But [that is impossible]. As I have already said before, it is clear that there must be at least as much [reality] in the cause as in the effect and that thus, since I am a thinking thing and have a certain idea of God within me, I must concede that whatever I finally designate as my own cause is also a thinking substance containing the idea of all the perfections I attribute to God. It is possible once again to ask whether that cause originates from itself or from something else. If it comes from itself, then, given what I have said, it is obvious that the cause itself is God. For clearly, if it derives its power of existing from itself, it also undoubtedly has the power of actually possess- 50 ing all the perfections whose idea it contains within itself, that is, all those that I think of as existing in God. But if it is produced from some other cause, then I ask once again in the same way whether this cause comes from itself or from some other cause, until I finally reach a final cause, which will be God.

For it is clear enough that this questioning cannot produce an infinite regress, particularly because the issue I am dealing with here is a matter not only of the cause which once produced me but also—and most importantly—of the cause which preserves me at the present time.

And I cannot assume that perhaps a number of partial causes came together to produce me and that from one of them I received the idea of one of the perfections I attribute to God and from another the idea of another perfection, so that all those perfections are indeed found somewhere in the universe, but they are not all joined together in a single being who is God. Quite the contrary, [for] the unity and simplicity—or the inseparability of all those things present in God—is one of the principal perfections which I recognize in Him. And surely the idea of this unity of all His perfections could not have been placed in me by any cause from which I did not acquire ideas of the other perfections as well, for no single cause could have made it possible for me to understand that those perfections were joined together and inseparable, unless at the same time it enabled me to recognize what those perfections were.

And finally, concerning my parents, even if everything I have ever believed about them is true, it is perfectly clear that they are not the ones who preserve me and that, to the extent that I am a thinking thing, there is no way they could have even made me. Instead they merely produced certain arrangements in the material substance which, as I have judged the matter, contains me—that is, contains my mind, for that is all I assume I am at 51 the moment. And thus the fact that my parents contributed to my existence provides no problem for my argument. Given all this, however, from the mere fact that I exist and that I have the idea of a supremely perfect being, or God, I must conclude that I have provided an extremely clear proof that God does, indeed, exist.

All that is left now is to examine how I have received that idea from God. For I have not derived it from the senses, and it has never come to me unexpectedly, as habitually occurs with the ideas of things I perceive with the senses, when those ideas of external substances impinge, or seem to impinge, on my sense organs. Nor is it something I just made up, for I am completely unable to remove anything from it or add anything to it. Thus, all that remains is that the idea is innate in me, just as the idea of myself is also innate in me.

And obviously it is not strange that God, when He created me, placed that idea within me, so that it would be, as it were, the mark of the master craftsman impressed in his own work—not that it is at all necessary for this mark to be different from the work itself. But the fact that God created me makes it highly believable that He made me in some way in His image and likeness, and that I perceive this likeness, which contains the idea of God, by the same faculty with which I perceive myself. In other words, when I turn my mind's eye onto myself, I not only understand that I am an incomplete thing, dependent on something else, and one that aspires [constantly] to greater and better things without limit, but at the same time I also realize that the one I depend on contains within Himself all those greater things [to which I aspire], not merely indefinitely and potentially, but actually and infinitely, and thus that He is God. The entire force of my 52 argument rests on the fact that I recognize I could not possibly exist with the sort of nature I possess, namely, having the idea of God within me, unless God truly existed as well—that God, I say, whose idea is in me—the Being having all those perfections which I do not grasp but which I am somehow capable of touching in my thoughts, and who is entirely free of any defect. These reasons are enough to show that He cannot be a deceiver, for natural light clearly demonstrates that every fraud and deception depends upon some defect.

But before I examine this matter more carefully and at the same time look into other truths I could derive from it, I wish to pause here for a while to contemplate God himself, to ponder His attributes, and to consider, admire, and adore the beauty of His immense light, to the extent that the eyes of my darkened intellect can bear it. For just as we believe through faith that the supreme happiness of our life hereafter consists only in this contemplation of the Divine Majesty, so we know from experience that the same [contemplation] now, though far less perfect, is the greatest joy we are capable of in this life.

FOURTH MEDITATION

Concerning Truth and Falsity

In these last few days, I have grown accustomed to detaching my mind from my senses, and I have clearly noticed that, in fact, I perceive very little with any certainty about corporeal things and that I know a great deal 53 more about the human mind and even more about God. As a result, I now have no difficulty directing my thoughts away from things I [perceive with the senses or] imagine, and onto those purely intellectual matters divorced from all material substance. And clearly the idea I have of the human mind, to the extent that it is a thinking thing that has no extension in length, breadth, and depth and possesses nothing else which the body has, is much more distinct than my idea of any corporeal substance. Now, when I direct my attention to the fact that I have doubts, in other words, that [I am] something incomplete and dependent, the really clear and distinct idea of an independent and complete being, that is, of God, presents itself to me. From this one fact—that there is an idea like this in me—or else because of the fact that I, who possess this idea, exist, I draw the clear conclusion that God also exists and that my entire existence depends on Him every single moment [of my life]. Thus, I believe that the human intellect can know nothing with greater clarity and greater certainty. And now it seems to me I see a way by which I can go from this contemplation of the true God, in whom all the treasures of science and wisdom are hidden, to an understanding of everything else.

First of all, I recognize that it is impossible that God would ever deceive me, for one discovers some sort of imperfection in everything false or deceptive. And although it may appear that the ability to deceive is evidence of a certain cleverness or power, the wish to deceive undoubtedly demonstrates either malice or mental weakness, and is therefore not found in God.

Then, I know from experience that there is in me a certain faculty of judgment, which I certainly received from God, like all the other things within me. Since He is unwilling to deceive me, He obviously did not 54 give me the kind of faculty that could ever lead me into error, if I used it correctly.

There would remain no doubt about this, if it did not seem to lead to the conclusion that I could never make mistakes. For if whatever is within me

I have from God and if He did not give me any power to commit errors, it would appear that I could never make a mistake. Now, it is true that as long as I am thinking only about God and directing myself totally to Him, I detect no reason for errors or falsity. But after a while, when I turn back to myself, I know by experience that I am still subject to innumerable errors. When I seek out their cause, I notice that I can picture not only a certain real and positive [idea] of God, or of a supremely perfect being, but also, so to speak, a certain negative idea of nothingness, or of something removed as far as possible from every perfection, and [I recognize] that I am, as it were, something intermediate between God and nothingness—that is, that I am situated between a supreme being and non-being in such a way that, insofar as I was created by a supreme being, there is, in fact, nothing in me which would deceive me or lead me into error, but insofar as I also participate, to a certain extent, in nothingness or non-being—in other words, given that I myself am not a supreme being—I lack a great many things. Therefore, it is not strange that I am deceived. From this I understand that error, to the extent that it is error, is not something real which depends on God, but is merely a defect. Thus, for me to fall into error, it is not necessary that I have been given a specific power to do this by God. Instead, I happen to make mistakes because the power I have of judging what is true [and what is false], which I do have from God, is not infinite within me.

However, this is not yet entirely satisfactory, for error is not pure negation, but rather the privation or lack of a certain knowledge that somehow ought to be within me. But to anyone who thinks about the nature of God, it does not seem possible that He would place within me any power that is not a perfect example of its kind or that lacks some perfection it ought to have. For [if it is true] that the greater the skill of the craftsman, the more perfect the works he produces, what could the supreme maker of all things create which was not perfect in all its parts? And there is no doubt that God could have created me in such a way that I was never deceived, and, similarly, there is no doubt that He always wills what is best. So then, is it better for me to make mistakes or not to make them?

55

As I weigh these matters more attentively, it occurs to me, first, that I should not find it strange if I do not understand the reasons for some of the things God does; thus I should not entertain doubts about His existence just because I happen to learn from experience about certain other things and do not grasp why or how He has created them. For given the fact that I already know my nature is extremely infirm and limited and that, by contrast, the nature of God is immense, incomprehensible, and infinite, I understand sufficiently well that He is capable of innumerable things about whose causes I am ignorant. For that reason alone, I believe that the entire class of causes we are in the habit of searching out as *final causes*[27] is completely useless in matters of physics, for I do not think I am capable of investigating the final purposes of God without appearing foolhardy.

It also occurs to me that, whenever we look into whether the works of God are perfect, we should not examine one particular creature by itself, but rather the universal totality of things. For something which may well justly appear, by itself, very imperfect, is utterly perfect [if we think of 56 it] as part of the [entire] universe. And although, given my wish to doubt everything, I have up to now recognized nothing as certain, other than the existence of myself and God, nonetheless, since I have observed the immense power of God, I cannot deny that He may have created many other things or at least is capable of creating them and therefore that I may occupy a place in a universe of things.

After that, by examining myself more closely and looking into the nature of my errors (the only things testifying to some imperfection in me), I observe that they proceed from two causes working together simultaneously, namely, from the faculty of knowing, which I possess, and from the faculty of choosing, or from my freedom to choose—in other words from both the intellect and the will together. For through my intellect alone I [do not affirm or deny anything, but] simply grasp the ideas of things about which I can make a judgment, and, if I consider my intellect in precisely this way, I find nothing there which is, strictly speaking, an error. For although countless things may well exist of which I have no idea at all within me, I still should not assert that I am deprived of them, in the proper sense of that word, [as if that knowledge were something my understanding was entitled to thanks to its nature]. I can only make the negative claim that I do not have them, for obviously I can produce no reason which enables me to prove that God ought to have given me a greater power of understanding than He has provided. And although I know that a craftsman is an expert, still I do not assume that he must therefore place in each of his works all the perfections he is capable of placing in some. Moreover, I certainly cannot complain that I have received from God a will or a freedom to choose that is insufficiently ample and perfect. For I clearly know from experience that my will is not circumscribed by any limits. And what seems to me particularly worthy of notice is the fact that, apart from my will, there

27 The final cause of something is (roughly) the purpose or reason for that thing's existence: e.g., the final cause of a statue might be an original idea or artistic goal in the sculptor's head which prompted her to make that particular statue. This terminology goes back to Aristotle, and involves a contrast between final causes and three other sorts of cause—material causes (the marble out of which the statue is hewn), formal causes (the shape—the form—of the statue), and efficient causes (the sculptor's craft in making the statue).

57 is nothing in me so perfect or so great that I do not recognize that it could be still more perfect or even greater. For, to consider an example: if I think about the power of understanding, I see at once that in me it is very small and extremely limited. At the same time, I form an idea of another understanding which is much greater, even totally great and infinite, and from the mere fact that I can form this idea, I see that it pertains to the nature of God. By the same reasoning, if I examine my faculty of memory or of imagination or any other faculty, I find none at all which I do not recognize as tenuous and confined in me and immense in God. It is only my will or my freedom to choose which I experience as so great in me that I do not apprehend the idea of anything greater. Thus, through my will, more than through anything else, I understand that I bear a certain image of and resemblance to God. For although the will is incomparably greater in God than in myself—because the knowledge and power linked to it make it much stronger and more efficacious and because, with respect to its object, His will extends to more things—nonetheless, if I think of the will formally and precisely in and of itself, His does not appear greater than mine. For the power of will consists only in the ability to do or not to do [something] (that is, to affirm or to deny, to follow or to avoid)—or rather in this one thing alone, that whether we affirm or deny, follow or avoid [something] which our understanding has set before us, we act in such a way that we do not feel that any external force is determining what we do. For to be free, I do not have to be inclined in two [different] directions. On the contrary, the more I am inclined to one—whether that is because I understand that
58 principles of the true and the good are manifestly in it or because that is the way God has arranged the inner core of my thinking—the more freely I choose it. Clearly divine grace and natural knowledge never diminish liberty, but rather increase and strengthen it. However, the indifference I experience when there is no reason urging me to one side more than to the other is the lowest degree of liberty. It does not demonstrate any perfection in [the will], but rather a defect in my understanding or else a certain negation. For if I always clearly perceived what is true and good, I would never need to deliberate about what I ought to be judging or choosing, and thus, although I would be entirely free, I could never be indifferent.

For these reasons, however, I perceive that the power of willing, which I have from God, considered in itself, is not the source of my errors. For it is extremely ample and perfect. And the source is not my power of understanding. For when I understand something, I undoubtedly do so correctly, since my [power of] understanding comes from God, and thus it is impossible for it to deceive me. So from where do my errors arise? Surely from the single fact that my will ranges more widely than my intellect, and I do not keep it within the same limits but extend it even to those things which I do not understand. Since the will does not discriminate among these things, it easily turns away from the true and the good, and, in this way, I make mistakes and transgress.

For example, in the past few days, when I was examining whether anything in the world existed and I observed that, from the very fact that I was exploring this [question], it clearly followed that I existed, I was not able [to prevent myself] from judging that what I understood so clearly was true, not because I was forced to that conclusion by any external force, but 59 because a great light in my understanding was followed by a great inclination in my will, and thus the less I was indifferent to the issue, the more spontaneous and free was my belief. For example: now I know that I exist, to the extent I am a thinking thing; but I am in doubt about whether this thinking nature within me (rather, which I myself *am*) is of that corporeal nature also revealed to me. I assume that up to this point no reason has offered itself to my understanding which might convince me that I am, or am not, of corporeal nature. From this single fact it is clear that I am indifferent as to which of the two I should affirm or deny, or whether I should even make any judgment in the matter.

Furthermore, this indifference extends not merely to those things about which the understanding knows nothing at all, but also, in general, to everything which it does not recognize with sufficient clarity at the time when the will is deliberating about them. For, however probable the conjectures [may be] which draw me in one direction, the mere knowledge that they are only conjectures and not certain and indubitable reasons is enough to urge me to assent to the opposite view. In the past few days I have learned this well enough by experience, once I assumed that all those things I had previously accepted as absolutely true were utterly false, because of the single fact that I discovered they could in some way be doubted.

But when I do not perceive that something is true with sufficient clarity and distinctness, if, in fact, I abstain from rendering judgment, I am obviously acting correctly and am not deceived. But if at that time I affirm or deny, [then] I am not using my freedom to choose properly. If I make up 60 my mind [and affirm] something false, then, of course I will be deceived. On the other hand, if I embrace the alternative, then I may, indeed, hit upon the truth by chance, but that would not free me from blame, since natural light makes it clear that a perception of the understanding must always precede a determination of the will. And it is in this incorrect use of the freedom of the will that one finds the privation which constitutes the nature of error. Privation, I say, inheres in this act of the will, to the extent that it proceeds from me, but not in the faculty I have received from God, nor even in the act, insofar as it depends upon Him.

For I have no cause to complain at all about the fact that God has not given me a greater power of understanding or a more powerful natural light than He has, because it is in the nature of a finite intellect not to understand many things and it is in the nature of a created intellect to be finite. Instead, I should thank Him, who has never owed me anything, for His generosity, rather than thinking that He has deprived me of something He did not provide or else has taken it away.

And I also have no reason to complain on the ground that He gave me a will more extensive than my understanding. For since the will consists of only a single thing and is, so to speak, indivisible, it does not seem that its nature is such that anything could be removed [without destroying it]. And, of course, the more extensive my will, the more I ought to show gratitude to the one who gave it to me.

And finally I also ought not to complain because God concurs with me in bringing out those acts of will or those judgments in which I am deceived. For those actions are true and good in every way, to the extent that they depend on God, and in a certain way there is more perfection in me because I am capable of eliciting these actions than if I were not. But 61 privation, in which one finds the only formal reason for falsity and failure, has no need of God's concurrence, because it is not a thing, and if one links it to Him as its cause, one should not call it privation but merely negation. For obviously it is not an imperfection in God that He has given me freedom to assent or not to assent to certain things, when He has not placed a clear and distinct perception of them in my understanding. However, it is undoubtedly an imperfection in me that I do not use that liberty well and that I bring my judgment to bear on things which I do not properly understand. Nonetheless, I see that God could easily have created me so that I never made mistakes, even though I remained free and had a limited understanding. For example, He could have placed in my intellect a clear and distinct perception of everything about which I would ever deliberate, or He could have impressed on my memory that I should never make judgments about things which I did not understand clearly and distinctly, and done that so firmly that it would be impossible for me ever to forget. And I readily understand that, if God had made me that way, insofar as I have an idea of this totality, I would have been more perfect than I am now. But I cannot therefore deny that there may somehow be more perfection in this whole universe of things because some of its parts are not immune to errors and others are—more perfection than if all things were entirely alike. And I have no right to complain just because the part God wanted me to play in the universe is not the most important and most perfect of all.

Besides, even if I am unable to avoid errors in the first way [mentioned above], which depends upon a clear perception of all those things about which I need to deliberate, I can still use that other [method], which re- 62 quires me only to remember to abstain from rendering judgment every time the truth of something is not evident. For although experience teaches me that I have a weakness which renders me incapable of keeping [my mind] always focused on one and the same thought, I can still see to it that by attentive and frequently repeated meditation I remember that fact every time the occasion demands. In this way I will acquire the habit of not making mistakes.

Since the greatest and preeminent perfection of human beings consists in this ability to avoid mistakes, I think that with the discovery in today's meditation of the cause of error and falsity I have gained a considerable gift. Clearly the source of mistakes can be nothing other than what I have identified. For as long as I keep my will restrained when I deliver judgments, so that it extends itself only to those things which reveal themselves clearly and distinctly to my understanding, I will surely be incapable of making mistakes, because every clear and distinct perception is undoubtedly something [real]. Therefore, it cannot exist from nothing but necessarily has God as its author—God, I say, that supremely perfect being, who would contradict His nature if He were deceitful. And thus, [such a perception] is unquestionably true. I have learned today not only what I must avoid in order to ensure that I am never deceived, but also at the same time what I must do in order to reach the truth. For I will assuredly reach that if I only pay sufficient attention to all the things I understand perfectly and distinguish these from all the other things which I apprehend confusedly and obscurely. In future, I will pay careful attention to this matter.

63

FIFTH MEDITATION

Concerning the Essence of Material Things, and, Once Again, Concerning the Fact that God Exists

Many other [issues] concerning the attributes of God are still left for me to examine, [as well as] many things about myself, that is, about the nature of my mind. However, I will perhaps return to those at another time. Now (after I have taken note of what I must avoid and what I must do to arrive at the truth) nothing seems to be more pressing than for me to attempt to emerge from the doubts into which I have fallen in the last few days and to see whether I can know anything certain about material things.

But before I look into whether any such substances exist outside of me, I ought to consider the ideas of them, insofar as they are in my thinking, and see which of them are distinct and which confused.

For example, I distinctly imagine quantity (which philosophers commonly refer to as 'continuous' quantity)—that is, the length, breadth and depth of the quantity, or rather, of the object being quantified. Further, I enumerate the various parts of the object, and assign to those parts all sorts of sizes, shapes, locations, and local movements, and to those movements all sorts of durations.

And in this way I not only clearly observe and acquire knowledge of those things when I examine them in general, but later, by devoting my attention to them, I also perceive innumerable particular details about their shapes, number, motion, and so on, whose truth is so evident and so well 64 suited to my nature, that when I discover them for the first time, it seems that I am remembering what I used to know, rather than learning anything new, or else noticing for the first time things which were truly within me earlier, although I had not previously directed my mental gaze on them.

I believe that the most important issue for me to consider here is that I find within me countless ideas of certain things which, even if they perhaps do not exist outside of me at all, still cannot be called nothing. Although in a certain sense I can think of them whenever I wish, still I do not create them. They have their own true and immutable natures. For example, when I imagine a triangle whose particular shape perhaps does not exist and has never existed outside my thinking, it nevertheless has, in fact, a certain determinate nature or essence or form which is immutable and eternal, which I did not produce, and which does not depend upon my mind; this is clearly shown in the fact that I can demonstrate the various properties of that triangle, namely, that the sum of its three angles is equal to two right angles, that the triangle's longest side has its endpoints on the lines made by the triangle's largest angle, and so on. These properties I now recognize clearly whether I wish to or not, although earlier, when I imagined the triangle [for the first time], I was not thinking of them at all and therefore did not invent them.

In this case it is irrelevant if I tell [myself] that perhaps this idea of a triangle came to me from external things through my sense organs, on the ground that I have certainly now and then seen objects possessing a triangular shape. For I am able to think up countless other shapes about which there can be no suspicion that they ever flowed into me through my senses, 65 and yet [I can] demonstrate various properties about them, no less than I can about the triangle. All these properties are *something* and not pure nothingness, since I conceive of them clearly and distinctly, and, as I have shown above, thus they must be true. Besides, even if I had not proved this, the nature of my mind is certainly such that I cannot refuse to assent to them, at least for as long as I am perceiving them clearly. And I remember that, even in those earlier days, when I was attracted as strongly as possible to objects of sense experience, I always maintained that the most certain things of all were those kinds of truth which I recognized clearly as shapes, numbers, or other things pertinent to arithmetic or geometry or to pure and abstract mathematics generally.

But if it follows from the mere fact that I can draw the idea of some object from my thinking that all things which I perceive clearly and distinctly as pertaining to that object really do belong to it, can I not also derive from this an argument which proves that God exists? For clearly I find the idea of Him, that is, of a supremely perfect being, within me just as much as I do the idea of some shape or number. I know that [actual and] eternal existence belongs to His nature just as clearly and distinctly as [I know] that what I prove about some shape or number also belongs to the nature of that shape or number. And therefore, even if all the things I have meditated on in the preceding days were not true, for me the existence of God ought to have at least the same degree of certainty as [I have recognized] up to this point in the truths of mathematics. 66

At first glance, however, this argument does not look entirely logical but [appears to] contain some sort of sophistry.[28] For, since in all other matters I have been accustomed to distinguish existence from essence, I

28 That is, clever-sounding but deceptive reasoning.

can easily persuade myself that [existence] can also be separated from the essence of God and thus that I [can] think of God as not actually existing. However, when I think about this more carefully, it becomes clear that one cannot separate existence from the essence of God, any more than one can separate the fact that the sum of the three angles in a triangle is equal to two right angles from the essence of a triangle, or separate the idea of a valley from the idea of a mountain. Thus, it is no less contradictory to think of a God (that is, of a supremely perfect being) who lacks existence (that is, who lacks a certain perfection) than it is to think of a mountain without a valley.[29]

Nonetheless, although I cannot conceive of God other than as something with existence, any more than I can of a mountain without a valley, the truth is that just because I think of a mountain with a valley, it does not therefore follow that there is any mountain in the world. In the same way, just because I think of God as having existence, it does not seem to follow that God therefore exists. For my thinking imposes no necessity on things, and in the same way as I can imagine a horse with wings, even though no horse has wings, so I could perhaps attribute existence to God, even though no God exists.

But this [objection] conceals a fallacy. For from the fact that I cannot think of a mountain without a valley, it does not follow that a mountain and 67 valley exist anywhere, but merely that the mountain and valley, whether they exist or not, cannot be separated from each other. However, from the fact that I cannot think of God without existence, it does follow that existence is inseparable from God, and thus that He truly does exist. Not that my thought brings this about or imposes any necessity on anything, but rather, by contrast, because the necessity of the thing itself, that is, of the existence of God, determines that I must think this way. For I am not free to think of God without existence (that is, of a supremely perfect being lacking a supreme perfection) in the same way that I am free to imagine a horse with wings or without them.

Suppose somebody objects: Agreed that once one has assumed that God has every perfection it is in fact necessary to admit that He exists (because existence is part of perfection), but it is not necessary to make that assumption, just as it is unnecessary to assume that all quadrilaterals [can] be inscribed in a circle. For if one assumed that, one would have to conclude that any rhombus could be inscribed in a circle—but this is clearly false.[30] But this objection is invalid. For although it may not be necessary

---

29 That is, an upslope without a downslope.

30 Quadrilaterals are four-sided figures. A figure can be inscribed in a circle when a circle can be drawn that passes through each corner. Rhombuses are figures with four sides of equal length. Squares (a type of rhombus) can be inscribed

for me ever to entertain any thought of God, nevertheless, whenever I do happen to think of a first and supreme being, and, as it were, to derive an idea of Him from the storehouse of my mind, I have to attribute to Him all perfections, even though I do not enumerate them all at that time or attend to each one of them individually. And this necessity is obviously sufficient to make me conclude correctly, once I have recognized that existence is a perfection, that a first and supreme being exists. In the same way, it is not necessary that I ever imagine any triangle, but every time I wish to consider a rectilinear[31] figure with only three angles, I have to attribute to it those [properties] from which I correctly infer that its three angles are no 68 greater than two right angles, although at that time I may not notice this. But when I think about which figures [are capable of being] inscribed in a circle, it is not at all necessary that I believe every quadrilateral is included in their number. On the contrary, I cannot even imagine anything like that, as long as I do not wish to admit anything unless I understand it clearly and distinctly. Thus, there is a great difference between false assumptions of this kind and the true ideas which are innate in me, of which the first and most important is the idea of God. For, in fact, I understand in many ways that this [idea] is not something made up which depends upon my thought but [is] the image of a true and immutable nature: first, because I cannot think of any other thing whose essence includes existence, other than God alone; second, because I am unable to conceive of two or more Gods of this sort, and because, given that I have already assumed that one God exists, I see clearly that it is necessary that He has previously existed from [all] eternity and will continue [to exist] for all eternity; and finally because I perceive many other things in God, none of which I can remove or change.

But, in fact, no matter what reasoning I finally use by way of proof, I always come back to the point that the only things I find entirely persuasive are those I perceive clearly and distinctly. Among the things I perceive in this way, some are obvious to everyone, while others reveal themselves only to those who look into them more closely and investigate more diligently, but nevertheless once the latter have been discovered, they are considered no less certain than the former. For example, even though the fact that the hypotenuse of a right triangle is opposite the largest angle of the triangle is more apparent than the fact that the square of the hypotenuse is equal to the sum of the squares of the other two sides, nonetheless, after we have initially recognized the second fact, we are no less certain of its 69 truth [than we are of the other]. But where God is concerned, if I were not overwhelmed with prejudices, and if images of perceptible things were not

---

in a circle, but rhombuses not containing four right angles cannot.

31 Formed by straight lines.

laying siege to my thinking on all sides, there is certainly nothing I would recognize sooner or more easily than Him. For what is more inherently evident than that there is a supreme being; in other words, that God exists, for existence [necessarily and eternally] belongs to His essence alone?

And although it required careful reflection on my part to perceive this [truth], nonetheless I am now not only as sure about it as I am about all the other things which seem [to me] most certain, but also, I see that the certainty of everything else is so dependent on this very truth that without it nothing could ever be perfectly known.

For although my nature is such that, as long as I perceive something really clearly and distinctly, I am unable to deny that it is true, nevertheless, because I am also by nature incapable of always fixing my mental gaze on the same thing in order to perceive it clearly, [and because] my memory may often return to a judgment I have previously made at a time when I am not paying full attention to the reasons why I made such a judgment, other arguments can present themselves which, if I knew nothing about God, might easily drive me to abandon that opinion. Thus, I would never have any true and certain knowledge, but merely vague and changeable opinions. For example, when I consider the nature of a triangle, it is, in fact, very evident to me (given that I am well versed in the principles of 70 geometry) that its three angles are equal to two right angles, and, as long as I focus on the proof of this fact, it is impossible for me not to believe that it is true. But as soon as I turn my mental gaze away from that, although I still remember I perceived it very clearly, it could still easily happen that I doubt whether it is true, if, in fact, I had no knowledge of God. For I can convince myself that nature created me in such a way that I am sometimes deceived by those things I think I perceive as clearly as possible, especially when I remember that I have often considered many things true and certain that I later judged to be false, once other reasons had persuaded me.

However, after I perceived that God exists, because at the same time I also realized that all other things depend on Him and that He is not a deceiver, I therefore concluded that everything I perceive clearly and distinctly is necessarily true. Thus, even if I am not fully attending to the reasons why I have judged that something is true, if I only remember that I have perceived it clearly and distinctly, no opposing argument can present itself that would force me to have doubts. Instead, I possess true and certain knowledge about it—and not just about that, but about all other matters which I remember having demonstrated at any time, for example, [about the truths] of geometry and the like. For what argument could I now bring against them? What about the fact that I am created in such a manner that I often make mistakes? But now I know that I cannot be deceived about those things which I understand clearly. What about the fact that I used to consider many other things true and certain which I later discovered to be false? But I was not perceiving any of these [things] clearly and distinctly, and, in my ignorance of this rule [for confirming] the truth, I happened to believe them for other reasons which I later discovered to be less firm. What then will I say? Perhaps I am dreaming (an objection I recently made to myself), or else everything I am now thinking is no more true than what happens when I am asleep? But even this does not change anything: for surely even though I am asleep, if what is in my intellect is 71 clear, then it is absolutely true.

In this way I fully recognize that all certainty and truth in science depend only on a knowledge of the true God, so much so that, before I knew Him, I could have no perfect knowledge of anything else. But now I am able to understand innumerable things completely and clearly, about both God Himself and other intellectual matters, as well as about all those things in corporeal nature that are objects of study in pure mathematics.

SIXTH MEDITATION

Concerning the Existence of Material Things and the Real Distinction between Mind and Body

It remains for me to examine whether material things exist. At the moment, I do, in fact, know that they *could* exist, at least insofar as they are objects of pure mathematics, since I perceive them clearly and distinctly. For there is no doubt that God is capable of producing everything which I am capable of perceiving in this way, and I have never judged that there is anything He cannot create, except in those cases where there might be a contradiction in my clear perception of it. Moreover, from my faculty of imagination, which I have learned by experience I use when I turn my at-
72 tention to material substances, it seems to follow that they exist. For when I consider carefully what the imagination is, it seems nothing other than a certain application of my cognitive faculty to an object which is immediately present to it and which therefore exists.

In order to clarify this matter fully, I will first examine the difference between imagination and pure understanding. For example, when I imagine a triangle, not only do I understand that it is a shape composed of three lines, but at the same time I also see those three lines as if they were, so to speak, present to my mind's eye. This is what I call imagining. However, if I wish to think about a chiliagon, even though I understand that it is a figure consisting of one thousand sides just as well as I understand that a triangle is a figure consisting of three sides, I do not imagine those thousand sides in the same way, nor do I see [them], as it were, in front of me. And although, thanks to my habit of always imagining something whenever I think of a corporeal substance, it may happen that [in thinking of a chiliagon] I create for myself a confused picture of some shape, nevertheless, it is obviously not a chiliagon, because it is no different from the shape I would also picture to myself if I were thinking of a myriagon[32] or of any other figure with many sides. And that shape is no help at all in recognizing those properties which distinguish the chiliagon from other

32 A myriagon is a 10,000-sided polygon.

polygons. However, if it is a question of a pentagon, I can certainly understand its shape just as [well as] I can the shape of a chiliagon, without the assistance of my imagination. But, of course, I can also imagine the pentagon by applying my mind's eye to its five sides and to the area they contain. From this I clearly recognize that, in order to imagine things, I need a certain special mental effort that I do not use to understand them, 73 and this new mental effort reveals clearly the difference between imagination and pure understanding.

Furthermore, I notice that this power of imagining, which exists within me, insofar as it differs from the power of understanding, is not a necessary part of my own essence, that is, of my mind. For even if I did not have it, I would still undoubtedly remain the same person I am now. From this it would seem to follow that my imagination depends upon something different from [my mind]. I understand the following easily enough: If a certain body—my body—exists, and my mind is connected to it in such a way that whenever my mind so wishes it can direct itself (so to speak) to examine that body, then thanks to this particular body it would be possible for me to imagine corporeal things. Thus, the only difference between imagination and pure understanding would be this: the mind, while it is understanding, in some way turns its attention to itself and considers one of the ideas present in itself, but when it is imagining, it turns its attention to the body and sees something in it which conforms to an idea which it has either conceived by itself or perceived with the senses. I readily understand, as I have said, that the imagination *could* be formed in this way, if the body exists, and because I can think of no other equally convenient way of explaining it, I infer from this that the body probably exists—but only probably—and although I am looking into everything carefully, I still do not yet see how from this distinct idea of corporeal nature which I find in my imagination I can derive any argument which necessarily concludes that anything corporeal exists.

However, I am in the habit of imagining many things apart from the 74 corporeal nature which is the object of study in pure mathematics, such as colors, sounds, smells, pain, and things like that, although not so distinctly. And since I perceive these better with my senses, through which, with the help of my memory, they appear to have reached my imagination, then in order to deal with them in a more appropriate manner, I ought to consider the senses at the same time as well and see whether those things which I perceive by this method of thinking, which I call sensation, will enable me to establish some credible argument to prove the existence of corporeal things.

First of all, I will review in my mind the things that I previously believed to be true, because I perceived them with my senses, along with the reasons for those beliefs. Then I will also assess the reasons why I later called them into doubt. And finally I will consider what I ought to believe about them now.

To begin with, then, I sensed that I had a head, hands, feet, and other limbs making up that body which I looked on as if it were a part of me or perhaps even my totality. I sensed that this body moved around among many other bodies which could affect it in different ways, either agreeably or disagreeably. I judged which ones were agreeable by a certain feeling of pleasure and which ones were disagreeable by a feeling of pain. Apart from pain and pleasure, I also felt inside me sensations of hunger, thirst, and other appetites of this kind, as well as certain physical inclinations

75 towards joy, sadness, anger, and other similar emotions. And outside myself, besides the extension, shapes, and motions of bodies, I also had sensations of their hardness, heat, and other tactile qualities and, in addition, of light, colors, smells, tastes, and sounds. From the variety of these, I distinguished sky, land, sea, and other bodies, one after another. And because of the ideas of all those qualities which presented themselves to my thinking, although I kept sensing these as merely my own personal and immediate ideas, I reasonably believed that I was perceiving certain objects entirely different from my thinking, that is, bodies from which these ideas proceeded. For experience taught me that these ideas reached me without my consent, so that I was unable to sense any object, even if I wanted to, unless it was present to my organs of sense, and I was unable not to sense it when it was present. And since the ideas I perceived with my senses were much more vivid, lively, and sharp, and even, in their own way, more distinct than any of those which I myself intentionally and deliberately shaped by meditation or which I noticed impressed on my memory, it did not seem possible that they could have proceeded from myself. Thus, the only conclusion left was that they had come from some other things. Because I had no conception of these objects other than what I derived from those ideas themselves, the only thought my mind could entertain was that [the objects] were similar to [the ideas they produced]. And since I also remembered that earlier I had used my senses rather than my reason and realized that the ideas which I myself formed were not as vivid, lively, and sharp as those which I perceived with my senses and that most of the former were composed of parts of the latter, I easily convinced myself that I had nothing at all in my intellect which I had not previously had in my senses. I also maintained, not without reason, that this body, which, by some special right, I called my own, belonged to me more than

76 any other object, for I could never separate myself from it, as I could from other [bodies], I felt every appetite and emotion in it and because of it, and finally, I noticed pain and the titillation of pleasure in its parts, but not in any objects placed outside it. But why a certain strange sadness of spirit follows a sensation of pain and a certain joy follows from a sensation of

[pleasurable] titillation, or why some sort of twitching in the stomach, which I call hunger, is urging me to eat food, while the dryness of my throat [is urging me] to drink, and so on—for that I had no logical explanation, other than that these were things I had learned from nature. For there is clearly no relationship (at least, none I can understand) between that twitching [in the stomach] and the desire to consume food, or between the sensation of something causing pain and the awareness of sorrow arising from that feeling. But it seemed to me that all the other judgments I made about objects of sense experience I had learned from nature. For I had convinced myself that that was how things happened, before I thought about any arguments which might prove it.

However, many later experiences have gradually weakened the entire faith I used to have in the senses. For, now and then, towers which seemed round from a distance appeared square from near at hand, immense statues standing on the tower summits did not seem large when I viewed them from the ground, and in countless other cases like these I discovered that my judgments were deceived in matters dealing with external senses. And not just with external [senses], but also with internal ones as well. For what could be more internal than pain? And yet I heard 77 that people whose legs or arms had been cut off sometimes still seemed to feel pain in the part of their body which they lacked. Thus, even though I were to feel pain in one of my limbs, I did not think I could be completely certain that it was the limb which caused my pain. To these reasons for doubting sense experience, I recently added two extremely general ones. First, there was nothing I ever thought I was sensing while awake that I could not also think I was sensing now and then while asleep, and since I do not believe that those things I appear to sense in my sleep come to me from objects placed outside me, I did not see why I should give more credit to those I appear to sense when I am awake. Second, because I was still ignorant—or at least was assuming I was ignorant—of the author of my being, there seemed to be nothing to prevent nature from constituting me in such a way that I would make mistakes, even in those matters which seemed to me most true. As for the reasons which had previously convinced me of the truth of what I apprehended with my senses, I had no difficulty refuting them. For since nature seemed to push me to accept many things which my reason opposed, I believed I should not place much trust in those things nature taught. And although perceptions of the senses did not depend upon my will, I did not believe that was reason enough for me to conclude that they must come from things different from myself, because there could well be some other faculty in me, even one I did not yet know, which produced them.

But now that I am starting to gain a better understanding of myself and of the author of my being, I do not, in fact, believe that I should rashly ac-

78 cept all those things I appear to possess from my senses, but, at the same time, [I do not think] I should call everything into doubt.

First, since I know that all those things I understand clearly and distinctly could have been created by God in a way that matches my conception of them, the fact that I can clearly and distinctly understand one thing, distinguishing it from something else, is sufficient to convince me that the two of them are different, because they can be separated from each other, at least by God. The power by which this [separation] takes place is irrelevant to my judgment that they are distinct. And therefore, given the mere fact that I know I exist and that, at the moment, I look upon my nature or essence as absolutely nothing other than that I am a thinking thing, I reasonably conclude that my essence consists of this single fact: I am a thinking thing. And although I may well possess (or rather, as I will state later, although I certainly do possess) a body which is very closely joined to me, nonetheless, because, on the one hand, I have a clear and distinct idea of myself, insofar as I am merely a thinking thing, without extension, and, on the other hand, [I have] a distinct idea of body, insofar as it is merely an extended thing which does not think, it is certain that my mind is completely distinct from my body and can exist without it.

Moreover, I discover in myself faculties for certain special forms of thinking, namely, the faculties of imagining and feeling. I can conceive of myself clearly and distinctly as a complete being without these, but I cannot do the reverse and think of these faculties without me, that is, without an intelligent substance to which they belong. For the formal conception of them includes some act of intellection by which I perceive that they are different from me, just as [shapes, movement, and the other] modes [or accidents of bodies are different] from the object [to which they belong]. I also 79 recognize certain other faculties [in me], like changing position, assuming various postures, and so on, which certainly cannot be conceived, any more than those previously mentioned, apart from some substance to which they belong, and therefore they, too, cannot exist without it. However, it is evident that these [faculties], if indeed they [truly] exist, must belong to some corporeal or extended substance, and not to any intelligent substance, since the clear and distinct conception of them obviously contains some [form of] extension, but no intellectual activity whatsoever. Now, it is, in fact, true that I do have a certain passive faculty of perception, that is, of receiving and recognizing ideas of sensible things. But I would be unable to use this power unless some active faculty existed, as well, either in me or in some other substance capable of producing or forming these ideas. But this [active faculty] clearly cannot exist within me, because it presupposes no intellectual activity at all, and because, without my cooperation and often even against my will, it produces those ideas. Therefore I am left to conclude that it exists in some substance different from me that must contain, either formally or eminently, all the reality objectively present in the ideas produced by that faculty (as I have just observed above).[33] This substance is either a body, that is, something with a corporeal nature which obviously contains formally everything objectively present in the ideas, or it must be God, or some other creature nobler than the body, one that contains [those same things] eminently. But since God is not a deceiver, it is very evident that He does not transmit these ideas to me from Himself directly or even through the intervention of some other creature in which their objective reality is contained, not formally but only eminently. For since he has given me no faculty whatsoever for recognizing such a source, but by contrast, has endowed me with a powerful tendency to believe that these ideas are sent out from corporeal things, I do not see how it would be pos- 80 sible not to think of Him as a deceiver, if these [ideas] were sent from any source other than corporeal things. And therefore corporeal things exist. However, perhaps they do not all exist precisely in the ways I grasp them with my senses, since what I comprehend with my senses is very obscure and confused in many things. But at least [I should accept as true] all those things in them which I understand clearly and distinctly, that is, generally speaking, everything which is included as an object in pure mathematics.

But regarding other material things which are either merely particular, for example that the sun is of such and such a magnitude and shape, and so on, or less clearly understood, for example light, sound, pain, and things like that, although these may be extremely doubtful and uncertain, nonetheless, because of the very fact that God is not a deceiver and thus it is impossible for there to be any falsity in my opinions which I cannot correct with another faculty God has given me, I have the sure hope that I can reach the truth even in these matters. And clearly there is no doubt that all those things I learn from nature contain some truth. For by the term *nature*, generally speaking, I understand nothing other than either God himself or the coordinated structure of created things established by God, and by the term *my nature*, in particular, nothing other than the combination of all those things I have been endowed with by God.

However, there is nothing that nature teaches me more emphatically than the fact that I have a body, which does badly when I feel pain, which needs food or drink when I suffer from hunger or thirst, and so on. And therefore I should not doubt that there is some truth in this.

For through these feelings of pain, hunger, thirst, and so on, nature teaches me that I am not merely present in my body in the same way a 81 sailor is present onboard a ship, but that I am bound up very closely and, so

---

33 See the introduction to the *Meditations* for an explanation of the distinction between formally and objectively present, and note 24 in the Third Meditation for what it is to be eminently present.

to speak, mixed in with it, so that my body and I form a certain unity. For if that were not the case, then when my body was injured, I, who am merely a thinking thing, would not feel any pain because of it; instead, I would perceive the wound purely with my intellect, just as a sailor notices with his eyes if something is broken on his ship. And when my body needed food or drink, I would understand that clearly and not have confused feelings of hunger and thirst. For those sensations of thirst, hunger, pain, and so on are really nothing other than certain confused ways of thinking, which arise from the union and, as it were, the mixture of the mind with the body.

Moreover, nature also teaches me that various other bodies exist around my own and that I should pursue some of these and stay away from others. And certainly from the fact that I sense a wide diversity of colors, sounds, odors, tastes, heat, hardness, and similar things, I reasonably conclude that in the bodies from which these different sense perceptions come there are certain variations which correspond to these perceptions, even if they are perhaps not like them. And given the fact that I find some of these sense perceptions pleasant and others unpleasant, it is entirely certain that my body, or rather my totality, since I am composed of body and mind, can be affected by various agreeable and disagreeable bodies surrounding it.

However, many other things which I seemed to have learned from nature I have not really received from her, but rather from a certain habit I 82 have of accepting careless judgments [about things]. And thus it could easily be the case that these judgments are false—for example, [the opinion I have] that all space in which nothing at all happens to stimulate my senses is a vacuum, that in a warm substance there is something completely similar to the idea of heat which is in me, that in a white or green [substance] there is the same whiteness or greenness which I sense, that in [something] bitter or sweet there is the same taste as I sense, and so on, that stars and towers and anything else some distance away have bodies with the same size and shape as the ones they present to my senses, and things of that sort. But in order to ensure that what I perceive in this matter is sufficiently distinct, I should define more accurately what it is precisely that I mean when I say I have learned something from nature. For here I am taking the word *nature* in a more restricted sense than *the combination of all those things which have been bestowed on me by God*. For this combination contains many things which pertain only to the mind, such as the fact that I perceive that what has been done cannot be undone, and all the other things I grasp by my natural light [without the help of the body]. Such things are not under discussion here. This combination also refers to many things which concern only the body, like its tendency to move downward, and so on, which I am also not dealing with [here]. Instead, I am considering only those things which God has given me as a combination of mind and body. And so nature, in this sense, certainly teaches me to avoid those things which bring a sensation of pain and to pursue those which [bring] a sensation of pleasure, and such like, but, beyond that, it is not clear that with those sense perceptions nature teaches us that we can conclude anything about things placed outside of us without a previous examination by the understanding, because to know the truth about them seems to belong only to the mind and not to that combination [of body and mind]. And so, although a star does not make an impression on my eyes any greater than 83 the flame of a small candle, nonetheless, that fact does not incline me, in any real or positive way, to believe that the star is not larger [than the flame], but from the time of my youth I have made this judgment without any reason [to support it]. And although I feel heat when I come near the fire, and even pain if I get too close to it, that is really no reason to believe that there is something in the fire similar to that heat I feel, any more than there is something similar to the pain. The only thing [I can conclude] is that there is something in the fire, whatever it might be, which brings out in us those sensations of heat or pain. So, too, although in some space there is nothing which stimulates my senses, it does not therefore follow that the space contains no substances. But I see that in these and in a great many other matters, I have grown accustomed to undermine the order of nature, because, of course, these sense perceptions are, strictly speaking, given to me by nature merely to indicate to my mind which things are agreeable or disagreeable to that combination of which it is a part, and for that purpose they are sufficiently clear and distinct. But then I use them as if they were dependable rules for immediately recognizing the essence of bodies placed outside me. However, about such bodies they reveal nothing except what is confusing and obscure.

In an earlier section, I have already examined sufficiently why my judgments may happen to be defective, in spite of the goodness of God. However, a new difficulty crops up here concerning those very things which nature reveals to me as objects I should seek out or avoid, and also concerning the internal sensations, in which I appear to have discovered errors: for example, when someone, deceived by the pleasant taste of a certain food, eats a poison hidden within it [and thus makes a mistake]. Of course, in this situation, the person's nature urges him only to eat food which has a pleasant taste and not the poison, of which he has no 84 knowledge at all. And from this, the only conclusion I can draw is that my nature does not know everything. There is nothing astonishing about that, because a human being is a finite substance and thus is capable of only limited perfection.

However, we are frequently wrong even in those things which nature urges [us to seek]. For example, sick people are eager for drink or food which will harm them soon afterwards. One could perhaps claim that such people make mistakes because their nature has been corrupted. But this does not remove the difficulty, for a sick person is no less a true creature of God than a healthy one, and thus it seems no less contradictory that God has given the person a nature which deceives him. And just as a clock made out of wheels and weights observes all the laws of nature with the same accuracy when it is badly made and does not indicate the hours correctly as it does when it completely satisfies the wishes of the person who made it, in the same way, if I look on the human body as some kind of machine composed of bones, nerves, muscles, veins, blood, and skin, as if no mind existed in it, the body would still have all the same motions it now has in those movements that are not under the control of the will and that, therefore, do not proceed from the mind [but merely from the disposition of its organs]. I can readily acknowledge, for example, that in the case of a body sick with dropsy,[34] it would be quite natural for it to suffer from a parched throat, which usually conveys a sensation of thirst to the mind, and for its nerves and other parts also to move in such a way that it takes a drink and thus aggravates the illness. And when nothing like this is harming the body, it is equally natural for it to be stimulated by a similar dryness in the throat and to take a drink to benefit itself. Now, 85 when I consider the intended purpose of the clock, I could say that, since it does not indicate the time correctly, it is deviating from its own nature, and, in the same way, when I think of the machine of the human body as something formed for the motions which usually take place in it, I might believe that it, too, is deviating from its own nature, if its throat is dry when a drink does not benefit its own preservation. However, I am fully aware that this second meaning of the word *nature* is very different from the first. For it is merely a term that depends on my own thought, a designation with which I compare a sick person and a badly constructed clock with the idea of a healthy person and a properly constructed clock, and thus, the term is extrinsic to these objects. But by that [other use of the term *nature*] I understand something that is really found in things and that therefore contains a certain measure of the truth.

Now, when I consider a body suffering from dropsy, even though I say that its nature has been corrupted, because it has a dry throat and yet does not need to drink, clearly the word *nature* is merely an extraneous term. However, when I consider the composite, that is, the mind united with such a body, I am not dealing with what is simply a term but with a true error of nature, because this composite is thirsty when drinking will do it harm. And thus I still have to enquire here why the goodness of God does not prevent its nature, taken in this sense, from being deceitful.

At this point, then, my initial observation is that there is a great difference between the mind and the body, given that the body is, by its very

34 An abnormal accumulation of watery fluid in the body (now called edema).

nature, always divisible, whereas the mind is completely indivisible. For, in fact, when I think of [my mind], that is, when I think of myself as purely a thinking thing, I cannot distinguish any parts within me. Instead, I understand that I am something completely individual and unified. And although my entire mind seems to be united with my entire body, nonetheless, I know that if a foot or arm or any other part of the body is sliced off, that loss will not take anything from my mind. And I cannot call the faculties of willing, feeling, understanding, and so on parts of the mind because it is the same single mind that wishes, feels, and understands. By contrast, I cannot think of any corporeal or extended substance that my thought is not capable of dividing easily into parts. From this very fact, I understand that the substance is divisible. (This point alone would be enough to teach me that the mind is completely different from the body, if I did not already know that well enough from other sources.) 86

Furthermore, I notice that the mind is not immediately affected by all parts of the body, but only by the brain, or perhaps even by just one small part of it, namely, the one in which our *common sense*[35] is said to exist. Whenever this part is arranged in the same particular way, it delivers the same perception to the mind, even though the other parts of the body may be arranged quite differently at the time. This point has been demonstrated in countless experiments, which I need not review here.

In addition, I notice that the nature of my body is such that no part of it can be moved by any other part some distance away which cannot also be moved in the same manner by any other part lying between them, even though the more distant part does nothing. So, for example, in a rope ABCD [which is taut throughout], if I pull on part D at the end, then the movement of the first part, A, will be no different than it would be if I pulled at one of the intermediate points, B or C, while the last part, D, remained motionless. And for a similar reason, when I feel pain in my foot, physics teaches me that this sensation occurs thanks to nerves spread throughout the foot. These nerves stretch from there to the brain, like cords, and when they are pulled in my foot, they also pull the inner parts of the brain, where they originate, and stimulate in them a certain motion which nature has established to influence the mind with a sense of pain apparently present in the foot. However, since these nerves have to pass through the shin, the thigh, the loins, the back, and the neck in order to reach the brain from the foot, it can happen that, even if that portion of the nerves which is in the foot is not affected, but only one of the intermedi- 87

35 See note 19, Second Meditation. Descartes is probably thinking of the pineal gland here, a tiny structure located between the two hemispheres of the brain; this is because he knew it to be the only anatomical structure of the brain which existed as a single part, rather than one half of a pair.

ate portions, the motion created in the brain is exactly the same as the one created there by an injured foot. As a result, the mind will necessarily feel the identical pain. And we should assume that the same is true with any other sensation whatsoever.

Finally, I notice that, since each of those motions created in that part of the brain which immediately affects the mind introduces into it only one particular sensation, we can, given this fact, come up with no better explanation than that this sensation, out of all the ones which could be introduced, is the one which serves to protect human health as effectively and frequently as possible [when a person is completely healthy]. But experience testifies to the fact that all sensations nature has given us are like this, and thus we can discover nothing at all in them which does not bear witness to the power and benevolence of God. Thus, for example, when the nerves in the foot are moved violently and more than usual, their motion, passing through the spinal cord to the inner core of the brain, gives a signal there to the mind which makes it feel something—that is, it feels as if there is a pain in the foot. And that stimulates [the mind] to do everything it can to remove the cause of the pain as something injurious to the foot. Of course, God could have constituted the nature of human beings in such a way that this same motion in the brain communicated something else to the mind, for example, a sense of its own movements, either in the brain, or in the foot, or in any of the places in between—in short, of anything you wish. But nothing else would have served so well for the preservation of the body. In the same way, when we need a drink, a certain dryness arises in the throat which moves its nerves and, with their assistance, the inner parts of the brain. And this motion incites in the mind a sensation of thirst, because in this whole situation nothing is more useful for us to know than that we need a drink to preserve our health. The same is true for the other sensations.

From this it is clearly evident that, notwithstanding the immense goodness of God, human nature, given that it is composed of mind and body, cannot be anything other than something that occasionally deceives us. For if some cause, not in the foot, but in some other part through which the nerves stretch between the foot and the brain, or even in the brain itself, stimulates exactly the same motion as that which is normally aroused when a foot is injured, then pain will be felt as if it were in the foot, and the sensation will naturally be deceiving. Since that same motion in the brain is never capable of transmitting to the mind anything other than the identical sensation and since [the sensation] is habitually aroused much more frequently from an injury in the foot than from anything else in another place, it is quite reasonable that it should always transmit to the mind a pain in the foot rather than a pain in any other part of the body. And if sometimes dryness in the throat does not arise, as it usually does, from the fact that a drink is necessary for the health of the body, but from some different cause, as occurs in a patient suffering from dropsy, it is much better that it should deceive us in a case like that than if it were, by contrast, always deceiving us when the body is quite healthy. The same holds true with the other sensations.

This reflection is the greatest help, for it enables me not only to detect all the errors to which my nature is prone, but also to correct or to avoid them easily. For since I know that, in matters concerning what is beneficial to the body, all my senses show [me] what is true much more frequently than they deceive me, and since I can almost always use several of them to examine the same matter and, in addition, [can use] my memory, which connects present events with earlier ones, as well as my understanding, which has now ascertained all the causes of my errors, I should no longer fear that those things which present themselves to me every day through my senses are false. And I ought to dismiss all those exaggerated doubts of the past few days as ridiculous, particularly that most important [doubt] about sleep, which I did not distinguish from being awake. For now I notice a significant distinction between the two of them, given that our memory never links our dreams to all the other actions of our lives, as it [usually] does with those things which take place when we are awake. For clearly, if someone suddenly appears to me when I am awake and then immediately afterwards disappears, as happens in my dreams, so that I have no idea where he came from or where he went, I would reasonably judge that I had seen some apparition or phantom created in my brain [similar to the ones created when I am asleep], rather than a real person. But when certain things occur and I notice distinctly the place from which they came, 90 where they are, and when they appeared to me, and when I can link my perception of them to the rest of my life as a totality, without a break, then I am completely certain that this is taking place while I am awake and not in my sleep. And I should not have the slightest doubt about the truth of these perceptions if, after I have called upon all my senses, my memory, and my understanding to examine them, I find nothing in any of them which contradicts any of the others. For since God is not a deceiver, it must follow that in such cases I am not deceived. But because, in dealing with what we need to do, we cannot always take the time for such a scrupulous examination, we must concede that human life is often prone to error concerning particular things and that we need to acknowledge the frailty of our nature.

INDEX

From the Publisher

A name never says it all, but the word "Broadview" expresses a good deal of the philosophy behind our company. We are open to a broad range of academic approaches and political viewpoints. We pay attention to the broad impact book publishing and book printing has in the wider world; we began using recycled stock more than a decade ago, and for some years now we have used 100% recycled paper for most titles. Our publishing program is internationally oriented and broad-ranging. Our individual titles often appeal to a broad readership too; many are of interest as much to general readers as to academics and students.

Founded in 1985, Broadview remains a fully independent company owned by its shareholders—not an imprint or subsidiary of a larger multinational.

For the most accurate information on our books (including information on pricing, editions, and formats) please visit our website at www.broadviewpress.com. Our print books and ebooks are also available for sale on our site.

b

broadview press
www.broadviewpress.com

The interior of this book is printed on 100% recycled paper.